Beginning Scientific English

Beginning Scientific English

D. E. Royds-Irmak

BOOK 2

Nelson

Also published by Nelson
John Swales: Writing Scientific English
D. E. Royds-Irmak: Beginning Scientific English Book 1
D. E. Royds-Irmak: Key to Exercises in Beginning Scientific English Books 1 and 2

Thomas Nelson and Sons Ltd
36 Park Street London W1
P.O. Box 27 Lusaka
P.O. Box 18123 Nairobi
P.O. Box 21149 Dar es Salaam
77 Coffee Street San Fernando Trinidad

Thomas Nelson (Australia) Ltd
19–39 Jeffcott Street
West Melbourne Victoria 3003

Thomas Nelson and Sons (Canada) Ltd
81 Curlew Drive Don Mills Ontario

Thomas Nelson (Nigeria) Ltd
P.O. Box 336 Apapa Lagos

First published 1975

17 555122 7

Printed in Great Britain by
Hazell Watson & Viney Ltd, Aylesbury, Bucks

Contents

Preface

Understanding and using scientific English presents many stumbling blocks to those who have achieved a reasonable proficiency in conversational or literary language through the usual course books. The present volume is designed, therefore to help bridge the gap between such a course-book and the science text-book.

Presenting its elementary science material in 40 short, graded texts, 1–22 in Book 1 and the remainder in Book 2, together with Exercises on vocabulary and language items which take a prominent place in technical writing, this book will, I hope, help students of English to overcome the bewildering problems of unfamiliar subject matter, new vocabulary and structures, and the new concept of English language used, not to relate personal experiences or to tell stories, but as an informative, impersonal account of principles and phenomena.

This book may be used by either the English-language teacher or by the Science-in-English teacher, who often justifiably complains that the teaching of his subject is greatly hampered by the students' inability to cope with language difficulties.

Work which can be dealt with only in small groups of students, or which requires equipment beyond the means of the ordinary classroom, has been avoided. It also seems essential, in accordance with the principles of learning through meaningful association, that work on language should remain within the limits imposed by the Texts themselves. Therefore, the exercises do not contain any vocabulary or structural items extraneous to the Texts.

The layout in each Unit is as follows:

1 At the head of each Unit, vocabulary which can most efficiently be taught through mother-tongue equivalents.

2 Text A, which presents the subject matter in simple, everyday language and structure.

3 Text B, which contains exactly the same subject matter, but is reworded in more scientific terms and structures. Each numbered word or phrase corresponds with a word or phrase italicised in Text A, and these are set out for easy reference in Exercise 1 of each Unit.

4 Exercises, with brief explanations, pinpointing important language items introduced in Text B. Special emphasis is given to passive

structures, since these are particularly common in scientific language.

5 Comprehension questions designed to check that content and vocabulary have been mastered, and to give practice in handling them

6 Questions for further discussion, designed to stimulate thought and interest, to develop powers of deduction and application of the principles dealt with in the Texts, and to provide opportunities for the student to achieve greater fluency in free discussion along the lines suggested in the Notes for the Guidance of Teachers given at the end of the book.

Some of these questions require the student to do a little research or enquiry outside class, and can be developed into project work, but most need him only to draw on his general knowledge and observation of everyday life.

7 Where possible, suggestions for further activities are included at the end of the Unit, in order to stimulate interest and reinforce learning. They may be carried out in class, where time and space permit, or allocated to be done outside class by a student or a group of students. They have purposely been kept simple and safe, and require a minimum of equipment, such as can be found in the home, or acquired with little difficulty.

8 Revision Exercises are provided at intervals, and may be used for reviewing work done, or for testing purposes. In either case, students should not refer to the Units while doing them.

9 Finally, two word lists are given at the end of the book. The first Vocabulary alphabetically lists the vocabulary given at the head of each Text A together with the words given after Questions for further discussion, and space is provided for the mother-tongue equivalents. The second is an English-to-English Glossary of the numbered terms and phrases of Text B, together with their explanations. A word or phrase occurring in the Vocabulary List of Book 1 is not repeated in that of Book 2, but a language item is included in the Glossary of Book 2 whether or not it has already appeared in Book 1.

How to Use this Book

To the Teacher

It is advisable to work through the Texts and Exercises in the order in which they are presented, since the subject matter is graded and grouped in a logical sequence, and vocabulary is built up progressively from one Unit to the next.

Discretion may be used as to how much reading aloud is done, and which Exercises are done in class and which as homework. But it is recommended that your students should hear a Text before attempting to read it themselves, and that the Exercises should be at least started in class before the student is left to his own devices with them.

Notes for the Guidance of Teachers will obviously need to be simplified and enlarged upon when they are explained to the student.

If answering comprehension questions as oral classwork, quick short answers are time savers, but as written work, answers in complete sentences should be encouraged.

When the student is familiar with the presentation of Text B as a rewording of Text A, he will find it useful to try to 'read' Text B while looking only at Text A. This is a good way of testing himself, and pinpoints weaknesses or gaps in his performance. Although this should never be attempted until the whole Unit has been completed, it brings a real sense of achievement when done fluently.

The student may be encouraged to keep a Science Notebook, in which to record new vocabulary and useful diagrams and illustrations, to work the Exercises and to make notes on class discussions and on any research or enquiry he may pursue on his own, particularly on the work involved in the suggestions for further activities.

Suggestions for Working

1 Give mother-tongue equivalents for the vocabulary listed at the head of each Unit. (As a learning reinforcement, these can be filled in later in the space provided in the Vocabulary list at the end of the book.)
2 Read Text A while students follow.
3 Read aloud each numbered word or phrase in Text B, while students find its equivalent italicised in Text A.
4 Have the student (or group) repeat each numbered phrase in Text B, before it is read in its entirety.
5 Work through the Exercises, or use them as homework, leaving time to deal orally with the Questions for further discussion.
A Key to the Exercises is available.

Acknowledgements
The cover design is by Jacky Wedgwood. Cover picture (and picture on page 64) is by Kadir Kir. The illustrations are by Design Practitioners Limited.

Unit 23 Falling Bodies

Vocabulary
compact-ness force relative

A
When a body is *not held up*, it falls because it is *pulled* towards the centre of the earth by the force of gravity.

Any object *which is falling* freely through the air *speeds up* each second by *about* 32 feet *in every* second. In other words, the *speeding up/which is*
5 *made* by the force of gravity is 32 feet per second per second.

When a thing is going very fast indeed, the *push of the air against it/ slightly changes* this speeding up. *Really,* there is a *greatest speed* at which any object may fall freely through air, *no matter/for how long it falls.* This is known as its *end speed,* and it depends on the shape and relative *com-*
10 *pactness* of the falling body. *Air pushing against* an object *which is not very compact,/like* a football, *makes it fall* much more slowly than objects *which are compact,* like stones.

B

When a body is (1)unsupported, it falls because it is (2)attracted towards the centre of the earth by the force of gravity.

Any object (3)falling freely through the air (4)accelerates each second by (5)approximately 32 feet (6)per second. In other words, the (7)accelera- 5 tion (8)produced by the force of gravity is 32 feet (6)per second (6)per second.

(9)At high velocities, the (10)resistance of the air (11)modifies this (7)acceleration. (12)In fact, there is a (13)maximum velocity at which any object may fall freely through air, (14)regardless of (15)the duration 10 of its fall. This is known as its (16)terminal velocity, and it depends on the shape and relative (17)density of the falling body. (18)Air resistance to an object (19)of low density, (20)such as a football, (21)causes it to fall much more slowly than objects (22) of high density, (20)such as stones.

Exercise 1 Find the way in which the words and phrases italicised in Text A are expressed in Text B:

1 not hold up
2 pulled
3 which is falling
4 speeds up
5 about
6 in every
7 speeding up
8 which is made
9 When a thing is going very fast indeed
10 push of the air against it
11 slightly changes
12 Really
13 greatest speed
14 no matter
15 for how long it falls
16 end speed
17 compactness
18 Air pushing against
19 which is not very compact
20 like
21 makes it fall
22 which are compact

Exercise 2 IN-/UN- placed in front of an adjective/past participle makes it negative in meaning. Complete the following table as in the example given:

	MEANING	ADJ./PAST PART.	NEGATIVE
1	held up	supported	UNsupported
2	covered with a shiny metal		
3	made hot		
4	can be seen		
5	slightly changed		
6	thrown back		
7	taken in		
8	shiny		
9	taken out		
10	held inside		

Exercise 3 The WHICH + verb clauses are often shortened (contracted) in scientific writing:

(1a) An object *which falls* freely accelerates. (present active)
(1b) An object *falling* freely accelerates.
(2a) An object *which is dropped* from a height falls downwards. (passive)
(2b) An object *dropped* from a height falls downwards.

Shorten (contract) the italicised WHICH + verb clauses in these sentences:

1 Acceleration *which is produced* by the force of gravity is 32 feet per second per second.
2 Air resistance *which occurs* at high velocities modifies this acceleration.
3 An object *which falls* freely in a vacuum meets no air resistance.

4 The maximum velocity *which is reached* by a body *which falls* freely in air is known as its terminal velocity.
5 The terminal velocity *which is reached* by a body *which falls* freely in air depends on its shape and relative density.
6 An object of low density *which is dropped* from a height in air will accelerate less than one of high density.
7 Resistance *which causes* an object of low density to fall more slowly is produced by air.
8 An object *which is unsupported* by anything will fall towards the centre of the earth.

Nouns ending with -ion

These verbs all form nouns ending in -ION:

	VERB	NOUN
a	Add -ION:	
	to reflect	reflect-ion
	to convect	convect-ion
	to conduct	conduct-ion
	to extract	extract-ion
	to prevent	prevent-ion
	to attract	attract-ion

	VERB	NOUN
b	(i) Drop E, add -ION:	
	to radiate	radiat-ion
	to insulate	insulat-ion
	to penetrate	penetrat-ion
	to accelerate	accelerat-ion
	(ii) Drop E, add -TION:	
	to produce	produc-tion
	(iii) B to P, add -TION:	
	to absorb	absorp-tion

Exercise 4 Each of the above nouns can be used, once only, to complete one of these sentences:

1 Heat can be transferred in three ways: (i) by ——, (ii) by ——, and (iii) by ——.
2 In a vacuum flask, the vacuum between the glass walls is produced by the —— of air.
3 Good heat —— can be obtained using any material which encloses plenty of air.
4 In a vacuum flask, the —— of heat from the outside and the escape of heat from the interior is prevented by silvering the walls.
5 Silvering the interior walls of the flask causes the —— of heat back into the flask.
6 In a vacuum flask, heat is preserved for long periods through the —— of heat loss by all three methods of heat transfer.
7 —— of heat by a dull surface is greater than by a polished one.
8 The —— towards the centre of the earth is known as the force of gravity.

9 In bodies free-falling in air, the —— of terminal velocity depends on the duration of the fall.
10 The —— of falling bodies is modified by the resistance of the air.

Exercise 5 Another form of nouns can be made by adding -OR to some verbs: Something which *conducts* is called a *conductor*.

What do we call:

1 something which radiates?
2 something which insulates?
3 something which convects?
4 something which extracts?
5 something which accelerates?
6 something which directs?
7 something which converts?
8 something which reflects?

Exercise 6 Read and rewrite this summary, using more scientific terms in place of the words and phrases italicised:

A body which is *not held up* falls towards the centre of the earth because of the *pull* of the force of gravity. If it is falling freely, it *speeds up* each second by *about* 32 feet *in every* second. *The push of the air against it* will *slightly change* its *speeding up,* depending on its shape and relative *compactness.* But, *no matter how long it falls,* in air, it will reach its *greatest speed* because of air *pushing against it.*

Exercise 7 Answer these questions without referring to the Texts:

1 Why does an unsupported body fall?
2 When a body is falling freely in air, does it always travel at the same speed?
3 What is the rate of acceleration, approximately?
4 On what does the maximum velocity of a freely-falling body depend?
5 What is the maximum velocity known as?
6 What slows down the acceleration at high velocities?
7 Which would fall faster, a basketball or a stone the same size?

Exercise 8 Questions for further discussion:

1 (a) Which is heavier, a pound of feathers or a pound of lead?
(b) Which has a higher density?
(c) Which would hit the ground first if they were dropped simultaneously from an aeroplane?

2 When thrown, why does a paper aeroplane fly faster (and straighter) than a flat sheet of paper?
3 When would you feel the greater air resistance, when travelling on a bicycle or on a motor bicycle? Why?
4 Why are aeroplanes built with pointed noses?

Exercise 9 Suggestions for further activities:

Hold a flat sheet of paper horizontally in one hand and a piece of chalk in the other. Hold them at the same height from the ground and let them fall simultaneously. Which reaches the ground first? Why? Now crumple the sheet of paper into a tight ball. Again hold it and the piece of chalk at the same height, and let them fall simultaneously. Now which one reaches the ground first? Why? The weight of the paper has not changed, what makes the difference?

Unit 24 Nitrogen Fixation

Vocabulary

ammonia element nitrogen tissue
carbon

A

Most of the atmosphere *is made of* nitrogen, but the nitrogen which plants and animals *need* to build up their tissues is never *there ready to use* in *large enough amounts*. This is *because* they cannot use free nitrogen *to do this*. They can *take in and use* it only when it is combined with other
5 chemical elements.

However, the bodies of animals *cannot* make free nitrogen combine *like this*. *If there were no* bacteria in the soil, the world's supply of combined nitrogen would soon be all *used up*.

Some bacteria which are found in the roots of some *kinds* of plants fix
10 the free nitrogen *of the atmosphere*, and combine it with other elements, such as carbon, hydrogen and oxygen. Then plants use this combined nitrogen to build their tissues.

Some other bacteria which are found in the soil, combine the free nitrogen of the atmosphere with hydrogen *and make* ammonia, which
15 they use to form their own *living tissue*. When these bacteria die, the compounds *of nitrogen* remain in the soil and are there ready for other plants to use for their *feeding*.

B

(1)The greater part of the atmosphere (2)consists of nitrogen, but the nitrogen which is (3)required by plants and animals to build up their tissues is never (4)available in (5)sufficient quantities. This is (6)due to the fact that free nitrogen cannot be used (7)for this purpose. It can be
5 (8)assimilated only when it is combined with other chemical elements.

However, the bodies of animals (9)are unable to make free nitrogen combine (10) in this way. (11)If it were not for bacteria in the soil, the world's supply of combined nitrogen would soon be (12)exhausted.

(13)Certain bacteria which are found in the roots of (13)certain (14)varie-
10 ties of plants, fix the free (15)atmospheric nitrogen, and combine it with other elements, such as carbon, hydrogen and oxygen. Then this combined nitrogen is used by plants to build their tissues.

(13)Certain other bacteria which are found in the soil combine the free (15)atmospheric nitrogen with hydrogen (16)to make ammonia,
15 which they use to form their own (17)protoplasm. When these bacteria die, the (18)nitrogenous compounds remain in the soil and are (4)available for the (19)nutrition of other plants.

Exercise 1 Find the way in which the words and phrases italicised in Text A are expressed in Text B:

1 most
2 is made of
3 need
4 there ready to use
5 large enough amounts
6 because
7 to do this
8 take in and use
9 cannot
10 like this
11 If there were no
12 all used up
13 Some
14 kinds
15 of the atmosphere
16 and make
17 living tissue
18 of nitrogen
19 feeding

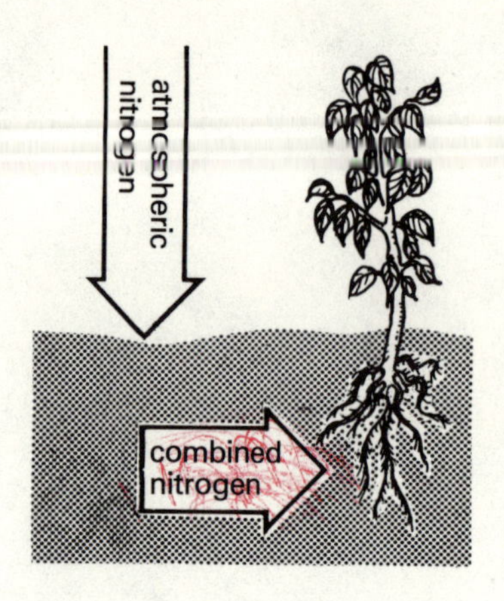

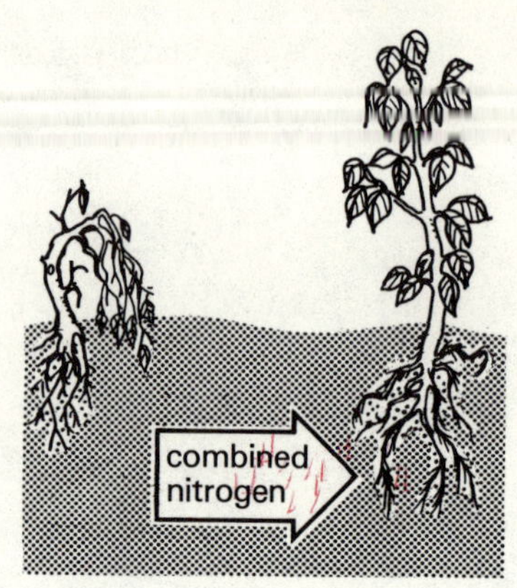

Exercise 2 In these pairs of sentences, one provides the reason for the fact expressed in the other. Decide which does this and join the two sentences, using DUE TO THE FACT THAT, as in example (c):

(a) Plants cannot use free nitrogen.
(b) A supply of combined nitrogen must be available.

Sentence (a) gives the reason for (b), so they are joined:

(c) A supply of combined nitrogen must be available, *due to the fact that* plants cannot use free nitrogen.

1 (a) Nitrogen can be assimilated only when combined.
(b) Plants are unable to use free nitrogen.

2 (a) The bacteria in the soil are of great importance.
(b) Without them the supply of combined nitrogen would soon be exhausted.

3 (a) There are certain kinds of bacteria in the soil which are able to combine free nitrogen.
(b) The world's supply of combined nitrogen is never exhausted.

4 (a) Nitrogenous compounds are available in the soil for the nutrition of plants.
(b) Certain bacteria have combined nitrogen, which remains in the soil after they die.

5 (a) These bacteria die and remain in the soil.
(b) The nitrogenous compounds which bacteria have formed can be used by plants.

6 (a) The candle burning under the bell-jar eventually went out.
(b) No oxygen remained to support combustion.

7 (a) Very little heat can escape from a vacuum flask.
(b) There is a vacuum between the glass walls, which are also silvered.

8 (a) At great heights above the earth's surface, the atmosphere is rarefied.
(b) Respiration is very difficult there.

Exercise 3 Rewrite these sentences, using IF IT WERE NOT FOR in place of 'if there were no', as in example (b):

(a) *If there were no* sun, life could not exist on earth.
(b) *If it were not for the* sun, life could not exist on earth.

1 If there were no bacteria in the soil, the supply of combined nitrogen would soon be exhausted.
2 If there were no dead bacteria in the soil, nitrogenous compounds would not be so plentiful there.
3 If there were no bacteria on the roots of certain varieties of plants, there would not be sufficient combined nitrogen.
4 If there were no vacuum between the walls of the flask, heat would escape through conduction or convection.
5 If there were no oxygen in the atmosphere, life could not exist on earth.
6 If there were no vegetation which died hundreds of millions of years ago, there would be no supplies of coal today.

Exercise 4 Rewrite this passage, using passive forms. You will then have summarised the Texts. The agent is required in each passive sentence. (The subjects of the passive sentences are italicised):

Plants and animals require *nitrogen* to build up their tissues, but they cannot use *free nitrogen* for this purpose. They can assimilate *it* only when it is combined. Certain bacteria fix *free atmospheric nitrogen* and combine *it* with other chemical elements, so that plants and animals can then assimilate *it*. Certain other bacteria combine *free nitrogen* and later other plants can use *these nitrogenous compounds* for their nutrition.

Exercise 5 Questions for further discussion:

1 What would you expect to find in nearly all fertilisers?
2 Why do farmers sometimes grow clover in a field and plough it back into the soil?
3 Why do waste materials from animals make good fertilisers?

Vocabulary

clover fertiliser

Unit 25 Light: (1) Lenses

Vocabulary

to collect | parallel | spot | upside down

A

A *glass which makes things look bigger/is useful in many ways.* Watchmakers, scientists and people *who cannot see near things well* use one. It is thicker in the middle than at the edges, and is known as a convex lens.

You can use a convex lens to *collect the rays of the sun on to one spot* and
5 *in this way* burn a hole in a piece of paper. The spot where the sun's rays are *collected* by the lens is known as the focal point. The distance between the focal point and the lens is known as the focal length. *Lenses of different thickness have different focal lengths.*

Another *kind* of lens is thinner in the middle than at the edges, and is
10 known as a concave lens. *Because* parallel light rays *which are passing* through a convex lens *bend towards each other*, this kind of lens is known as a converging lens. Because parallel light rays which are passing through a concave lens *bend away from each other*, this kind of lens is known as a diverging lens.

15 If you hold a *convex* lens in front of a window and hold a white card at the focal point, you will see a *picture* of the window on the card. The picture will be *upside down*, because the light rays from the window are made to bend towards each other by the convex lens.

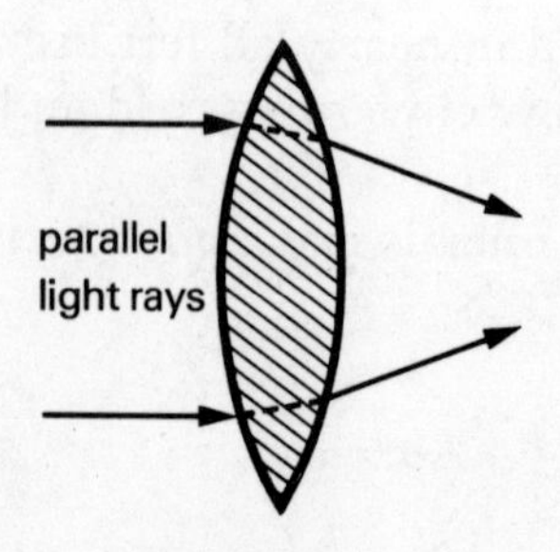

a. Convex lens

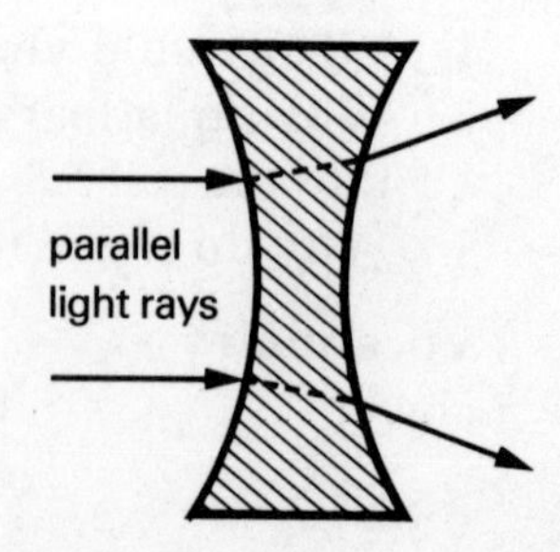

b. Concave lens

B

A (1)magnifying lens (2)has many uses. It is used by watchmakers, scientists and (3)long-sighted people. It is thicker in the middle than at the edges, and is known as a convex lens.

A convex lens can be used to (4)concentrate the sun's rays and (5)thus burn a hole in a piece of paper. The spot where the sun's rays are (6)concentrated by the lens is known as the focal point. The distance between the focal point and the lens is known as the focal length. (7)The focal length differs with the thickness of the lens.

Another (8)type of lens is thinner in the middle than at the edges, and is known as a concave lens. (9)Owing to the fact that parallel light rays (10)passing through a convex lens (11)converge, this (8)type of lens is known as a converging lens. (9)Owing to the fact that parallel light rays (10)passing through a concave lens (12)diverge, this (8)type of lens is known as a diverging lens.

If a (13)converging lens is held in front of a window and a white card is held at the focal point, an (14)image of the window will be seen on the card. The (14)image will be (15)inverted, because the light rays from the window are made to (11)converge by the (13)converging lens.

Exercise 1 Find the way in which the words and phrases italicised in Text A are expressed in Text B:

1 glass which makes things look bigger
2 is useful in many ways
3 who cannot see near things well
4 collect the rays of the sun onto one spot
5 in this way
6 collected
7 Lenses of different thickness have different focal lengths
8 kind
9 Because
10 which are passing
11 bend towards each other
12 bend away from each other
13 convex
14 picture
15 upside down

Exercise 2 In these pairs of sentences, one sentence provides the reason for the fact expressed in the other. Decide which sentence does this and join the two sentences, using OWING TO THE FACT THAT, as in example (c):

(a) Parallel light rays converge as they pass through.
(b) A convex lens is known as a converging lens.

Sentence (a) provides the reason for (b), so they are joined:

(c) A convex lens is known as a converging lens *owing to the fact that* parallel light rays converge as they pass through.

1 (a) Long-sighted people use a magnifying glass.
  (b) They cannot see near things clearly.
2 (a) Parallel light rays diverge as they pass through.
  (b) A concave lens is known as a diverging lens.
3 (a) The image of the window will be inverted.
  (b) The light rays from the window converge as they pass through the converging lens.
4 (a) A hole is burnt in a piece of paper.
  (b) The sun's rays are concentrated onto one spot.
5 (a) The focal length differs with the thickness of the lens.
  (b) The focal lengths of two lenses of different thickness will not be the same.

Exercise 3 Find ONE word to replace each phrase given below. Then use the words to complete the sentences correctly:

bend away from each other
collect onto one spot
upside down
who cannot see near things clearly
bend towards each other
made to look bigger
in this way

1 Objects are —— by convex lenses.
2 A magnifying glass is used by —— people.
3 Parallel light rays —— when they pass through a concave lens.
4 A converging lens will —— the sun's rays and —— burn a hole in a piece of paper.
5 The image of the window on the card will be —— because the rays of light will —— as they pass through the convex lens.

Exercise 4 Answer these questions without referring to the Texts. (Do not number your answers, but write them as a paragraph, giving all the answers in complete sentences.) You will then have summarised the Texts:

1 By what other names is a magnifying glass known?
2 What happens to parallel light rays passing through it?
3 How can a hole be burnt in a piece of paper?
4 What is (a) the focal point? and (b) the focal length?
5 What is a lens which is thinner in the middle known as?
6 What happens to parallel light rays passing through it?
7 How can an image of a window be seen on a white card?
8 Why will the image be inverted?
9 What causes the difference in focal length?

Exercise 5 Questions for further discussion:

1 Must lenses always be made of glass? What other materials could be used?
2 A plastic bag filled with water would make a converging lens. Is this true?
3 How can bottles, especially if filled with water, left on the ground cause a fire?
4 Could a glass of water left on a sunny window-sill start a fire? If so, how?

Exercise 6 Suggestions for further activities:

1 Get a magnifying glass and use it to concentrate the sun's rays onto a piece of white paper. Hold it very still for a time, and burn a hole in the paper. Now blacken a small area of the paper with a lead pencil, and repeat the experiment. Why do the sun's rays burn the blackened area more quickly?
2 Fill a clear plastic bag with water and fasten the mouth tightly. Can you use it as a magnifying glass?
3 Standing at a distance from a lighted window, take a magnifying glass to throw a clear image of the window onto a white card or plain wall. Ask someone to stand against the window and wave his arms above his head. This movement will show clearly that the image on the card is inverted.
4 Place a small piece of glass on a newspaper. Put one large drop of water or clear oil on to the glass. Does the writing under the drop look bigger? Why?

Unit 26 Light: (2) Reflection in mirrors

Vocabulary
exactly sphere

A

A mirror must be *flat* if it *is going to throw back* an image which is *exactly the same as* the *thing* in front. When you look at yourself in a flat mirror, you will see an *exact picture* of yourself, but *it has* one important difference. Your right side is on the left of the *picture in the mirror*, and your left side
5 is on the right of the picture in the mirror. *Or we can say*, the picture in the mirror is exactly the same, but *the sides are turned the other way*.

However, a mirror may not always be flat. It may have a curved surface. If the surface is concave, parallel light rays *which hit* it will *go towards each other* when they are *thrown back*. If the surface is convex, parallel
10 light rays will *go away from each other/when they are thrown back*.

A convex or a concave mirror may be part of a sphere. *If that is true*, it is known as a spherical mirror.

B

A mirror must be (1)plane if it (2)is to reflect an image which is (3)identical to the (4)object in front. When you look at yourself in a (1)plane mirror, you will see an (5)identical image of yourself, but (6)with one important difference. Your right side is on the left of the (7)mirror-image, and your
5 left side is on the right of the (7)mirror-image. (8)In other words, the image is (3)identical, but (9)laterally inverted.

However, a mirror may not always be (1)plane. It may have a curved surface. If the surface is concave, parallel light rays (10)striking it will (11)converge when they are (12)reflected. If the surface is convex,
10 parallel light rays will (13)diverge (14)when reflected.

A convex or a concave mirror may be part of a sphere. (15)In that case, it is known as a spherical mirror.

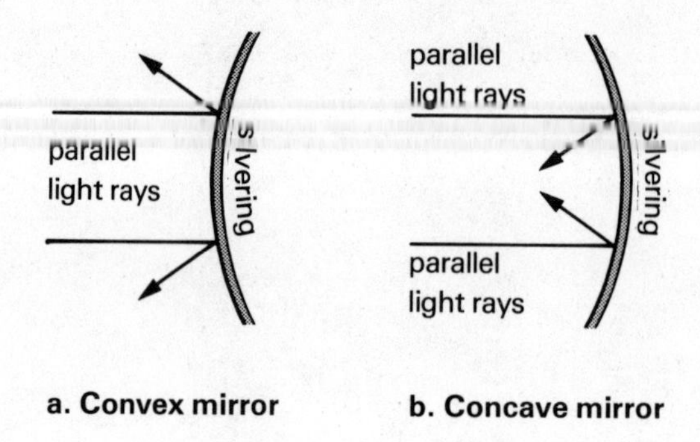

a. Convex mirror b. Concave mirror

Exercise 1 Find the way in which the words and phrases italicised in Text A are expressed in Text B:

1 flat
2 is going to throw back
3 exactly the same as
4 thing
5 exact picture
6 it has
7 picture in the mirror
8 Or we can say
9 the sides are turned the other way
10 which hit
11 go towards each other
12 thrown back
13 go away from each other
14 when they are thrown back
15 If that is true

Exercise 2 IN THAT CASE is frequently used instead of 'if this is true/when that happens':

(a) Silver is a good conductor of heat, and *if this is true*, it is a poor insulator.

(b) Silver is a good conductor of heat, and *in that case*, it is a poor insulator.

Rewrite these sentences, using IN THAT CASE, as in example (b):

1 A mirror may not be plane, and if this is true, the image will not be identical to the object.
2 A mirror may be concave, and if that is true, light rays will converge when reflected.
3 A mirror may be convex, and if that is true, light rays will diverge when reflected.
4 A mirror may be convex, and if that is true, the reflection will be smaller than the object.
5 A mirror may be concave, and if this is true, the object will be magnified.

6 A person may be long-sighted, and if this is true, he will need a magnifying glass to see near objects clearly.
7 Parallel light rays passing through a convex lens from the window converge, and when this happens, the image on the card will be inverted.
8 The sun's rays may be concentrated onto one spot through a converging lens, and when that happens, a hole is burnt in a piece of paper held at the focal point.
9 The air may be extracted from the bell-jar, and when that happens, the lighted candle goes out.
10 If no allowance is made for expansion, the rails will bend out of shape, and if that happens, railway accidents may occur.
11 A reservoir may be much higher than the town it supplies, and if that is true, gravity may provide the pressure to transport the water along the pipes.
12 Water expands on freezing, and if this is true, ice will float.
13 Boiling points of different substances differ, and if this is true, salt-water does not boil at 100°C.
14 Most green plants require sunlight, and if that is true, they cannot survive for long in the dark.

Exercise 3 After WHEN/IF the subject and 'to be' verb may be left out, if the subject is the same in both parts of the sentence:

(a) Parallel light rays converge *when they are reflected.*
(b) Parallel light rays converge *when reflected.*

Rewrite these sentences, shortening (contracting) the WHEN/IF clauses, as in example (b):

1 Parallel light rays will diverge when they are reflected from a convex surface.
2 When it is part of a sphere, a mirror is known as a spherical mirror.
3 If it is plane, a mirror reflects an identical but laterally inverted image.
4 If it is concave, the mirror reflects a magnified and laterally inverted image.
5 Parallel light rays will converge when they are reflected from the surface of a concave mirror.
6 When it is struck by parallel light rays, a plane mirror reflects a laterally inverted but identical image.
7 A card will show an inverted image if it is held at the focal point of a convex lens.

8 If the mirror is convex, it will reflect an image which is smaller than the object in front of it.
9 An image is always laterally inverted when it is seen as a mirror reflection.
10 An image is invariably inverted when it has been made by passing parallel light rays through a convex lens.

Exercise 4 Answer these questions without referring to the Texts:

1 Does a plane mirror reflect a magnified or an identical image?
2 What is the important difference between a plane mirror image and the object in front of it?
3 What do you understand by 'laterally inverted'?
4 What is the difference between being inverted and being laterally inverted?
5 Which sort of mirror will cause parallel light rays to (a) diverge and (b) converge when reflected?
6 What sort of lens will cause parallel light rays to (a) converge and (b) diverge when passing through?
7 What do you understand by 'spherical mirror'?

Exercise 5 Questions for further discussion:

1 What shape would a magnifying shaving mirror be?
2 What shape are most driving mirrors?
3 Have you ever seen mirrors in fun-fairs or Luna parks? What shapes are they, and what difference does the shape make to the reflection?
4 Why does a cheap mirror give a distorted (crooked) reflection, especially from a distance?

Vocabulary
fun-fair

Exercise 6 Suggestions for further activities:

1 Look at yourself in the bowl of a shiny spoon. Is your image inverted? Or magnified? Or laterally inverted?
2 Turn the spoon over. What differences are there in the reflected image now? Why?

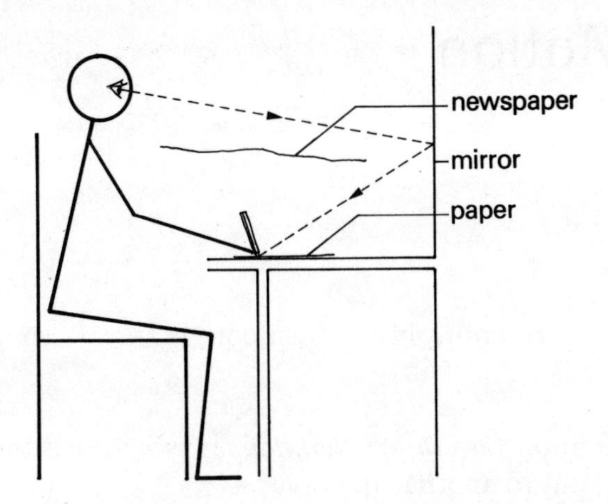

3 Seat yourself at a table opposite an upright mirror. Place a fairly large sheet of paper on the table in front of you, and take a pencil. Ask someone to hold a newspaper above the paper, so that you cannot see your own hand on it, but only the reflection of it in the mirror. Looking only at the reflection in the mirror, draw a square and its two diagonals. Why do you find it so difficult to do?

Unit 27 Wave Motion

Vocabulary

characteristic   to multiply   smooth   sound

A

One of the most important *things that happen in nature* is the *way energy is sent* from one point to another in waves.

When you drop a stone into a smooth lake, the surface is covered with *waves in circles* moving outwards from the centre point. The water itself does not *really* travel; it *only* rises and falls. *This kind of movement* is known as wave motion.

Light, heat and sound travel in waves, known as light waves, heat waves and sound waves.

We can measure three things in any wave. We can measure the distance from its highest to its lowest point, *that is*, amplitude. We can count the number of waves *coming in one second*, that is, frequency. We can measure the distance from the *highest point of one wave* to the highest point of the next, that is, wave-length.

We can also *do a sum to find out* the *speed* of any wave. We do this by multiplying frequency with wave-length, that is, *Speed = Frequency × Length/, when we write V for speed*, F for frequency and L for length.

A very large number of radio waves come in one second, and they travel at the speed of light. Different kinds of waves *have each the same* three characteristics: amplitude, frequency and length.

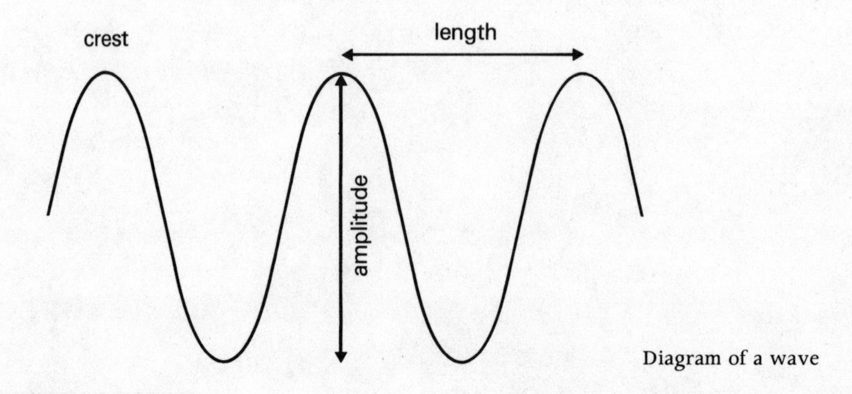

Diagram of a wave

B

One of the most important (1)natural phenomena is the (2)transmission of energy from one point to another in waves.

When a stone is dropped into a smooth lake, the surface is covered with (3)circular waves moving outwards from the centre point. The water
5 itself does not (4)in fact travel; it (5)merely rises and falls. (6)Motion of this kind is known as wave motion.

Light, heat and sound travel in waves, known as light waves, heat waves and sound waves.

In any wave, three things can be measured: amplitude, (7)i.e. the
10 distance from its highest to its lowest point; frequency, (7)i.e. the number of waves (8)per second; wave-length, (7)i.e. the distance from (9)one crest to the next.

We can also (10)calculate the (11)velocity of any wave, by multiplying frequency with wave-length, (7)i.e. (12)V=FL, (13)where V=velocity,
15 F=frequency and L=length.

(14)Radio waves possess a very high frequency, and travel at the (11)velocity of light. (6)Waves of different kinds (15)possess three characteristics (16)in common: amplitude, frequency and length.

Exercise 1 Find the way in which the words and phrases italicised in Text A are expressed in Text B:

1 things that happen in nature
2 way energy is sent on
3 waves in circles
4 really
5 only
6 This kind of movement
7 that is
8 coming in one second
9 highest point of one wave
10 do a sum to find out
11 speed
12 Speed = Frequency × Length
13 when we write V for speed
14 A very large number of radio waves come in one second
15 have
16 each the same

Exercise 2 Noun + OF THIS KIND is often used instead of 'this kind of' + noun:

In everyday speech or writing:
(a) *This kind of motion* is known as wave motion.

In scientific writing:
(b) *Motion of this kind* is known as wave motion.

Remember to put *a* or *an* where necessary:
(a) This type of problem. (b) A problem of this type.

Read or rewrite these sentences, changing the word order, as in example (b):

1 This kind of a space is known as a vacuum.
2 This sort of energy is known as chemical.
3 This type of bacteria can be found on the roots of certain plants.
4 This shape of jar is known as a bell-jar.
5 This length of a wave has a low frequency.
6 This type of mirror is known as spherical.
7 This sort of bacteria can fix free nitrogen.
8 This strength of pressure is not normal in the atmosphere.
9 This kind of attraction is known as gravitational.
10 This velocity of a wave will have a high frequency.
11 This frequency of a wave is known as electro-magnetic.
12 This variety of plant needs a hot, humid climate.
13 This amplitude of a wave will have a low frequency.
14 This type of experiment may be dangerous.
15 This sort of process is known as conduction.
16 This kind of current is known as a convection current.
17 This type of lens is called a converging lens.
18 This kind of dispersal is known as wind dispersal.
19 This type of water is known as soft.
20 This sort of rubber is known as synthetic.

Exercise 3 Rewrite this summary of the Texts, using a more scientific term for each phrase italicised. (One small *change in word order is necessary):

The *sending on* of energy in waves is an important *happening in nature*. When a stone is dropped into a lake, *waves *in circles* move outwards from the centre. The water is not in fact travelling, but merely *going up* and *going down*. This is known as wave motion. The *distance between the crest and the lowest point* of any wave can be measured. The *number of waves per second* can be determined. The *distance from the crest of one wave to the crest of the next* can also be measured. All waves have these three characteristics *each the same*. The velocity of any wave can be *worked out in a sum*, by multiplying frequency by length, i.e. V = FL (*when we write V for* velocity, and *FL for* frequency multiplied by length.)

Exercise 4 Answer these questions without referring to the Texts:

1 When a stone is dropped into a smooth lake, what is the surface covered with?
2 Does the moving water in fact travel?
3 What does it do?
4 What do we call motion of this kind?
5 What else, besides water, travels in waves?
6 How many characteristics have all waves in common?
7 How can we calculate the velocity of any wave?
8 Which sort of waves have a very high frequency?

Exercise 5 Questions for further discussion:

1 If the wind blew your hat out of reach into the middle of a small lake, could you get it back by making waves? (Give a reason for your answer.)
2 Waves of all varieties can be reflected. Give examples.

Exercise 6 Suggestions for further activities:

Nearly fill a wide bowl with water and float a match on the surface. Drop a small stone or bean-seed into the middle of the bowl and observe the small waves. Observe the movement of the match. Does it get carried to the side of the bowl by the waves?

Unit 28 Sound

Vocabulary

the blow	to disturb	lightning	to vibrate
brain	ear-drum	nerve	vibration
direction	hammering	thunder	

A

There can be no sound without movement. Sound is made by vibrations and travels in waves. The waves move from the vibrating object in every direction, in *circles which get bigger and bigger all the time*. They disturb the molecules in the *air round about* so that one molecule *hits* the next, *and so it continues*, until the vibration reaches the ear. *In the ear*, the ear-drum vibrates, the nerves *send messages* to the brain, *and the brain/ gives them a meaning* as sound.

If the *place where sound is coming from* vibrates in a vacuum, you cannot hear any sound. This is *because* there is no *surrounding material to use* (i.e. molecules) to *send* the sound to the ear.

The *speed* of sound is *changed* by the material used to send it. *This means* it travels slowest in air, faster in a liquid, and fastest in a solid.

The speed of sound is *very much* less than *the speed* of light. This is why we can see the lightning before we can hear the thunder, although they both occur *at exactly the same time*. It is also why, when we watch somebody hammering *far away*, we can see the hammer fall before we can hear the sound of the blow.

The speed of sound in air is *about* 1,100 feet or 330 metres *in every* second.

B

(1)No sound can exist without movement. Sound is made by vibrations and travels in waves. The waves move from the vibrating object in every direction, in (2)ever-increasing circles. They disturb the molecules in the (3)surrounding air, so that one molecule (4)strikes the next, (5)and so on, until the vibration reaches the ear. (6)There, the ear-drum vibrates, the nerves (7)transmit impulses to the brain, (8)which (9)interprets them as sound.

If the (10)source of sound vibrates in a vacuum, no sound can be heard. This is (11)due to the fact that there is no (12)medium (i.e. molecules) to (13)transmit the sound to the ear.

The (14) velocity of sound is (15)affected by the (12)medium which (13)transmits it. (16)That is to say, it travels slowest in air, faster in a liquid, and fastest in a solid.

The (14)velocity of sound is (17)considerably less than (18)that of light. This is why lightning can be seen before the thunder can be heard, although they both occur (19)simultaneously. It is also why, when we watch somebody hammering (20)at a distance, the hammer can be seen to fall before the sound of the blow can be heard.

The (14)velocity of sound in air is (21)approximately 1,100 feet or 330 metres (22)per second.

Exercise 1 Find the way in which the words and phrases italicised in Text A are expressed in Text B:

1 There can be no sound
2 circles which get bigger and bigger all the time
3 air round about
4 hits
5 and so it continues
6 In the ear
7 send messages
8 and the brain
9 gives them a meaning
10 place where sound is coming from
11 because
12 surrounding material to use
13 send
14 speed
15 changed
16 This means
17 very much
18 the speed
19 at exactly the same time
20 far away
21 about
22 in every

Exercise 2 Join each pair of sentences, using WHICH to avoid repeating the noun, as in example (b):

(a) Sound is made by vibrations. *The vibrations* reach the ear.
(b) Sound is made by vibrations *which* reach the ear.

1 The vibrations disturb the molecules. The molecules strike each other.
2 Sound is made by vibrations. The vibrations disturb the surrounding molecules.
3 The impulses are transmitted to the brain. The brain interprets them as sound.
4 The brain interprets the impulses. The impulses have been transmitted.
5 Sound travels through a medium. The medium affects its velocity.
6 The velocity of sound is affected by the medium. It travels through a medium.
7 We can see lightning before we can hear the thunder. The thunder occurs simultaneously.
8 The blow cannot be heard until after the hammer has been seen to fall. The blow occurs simultaneously.
9 The velocity of sound is approximately 330 metres per second. The velocity of sound is much less than that of light.
10 Sound travels fastest in a solid. Sound travels at approximately 330 metres per second in air.

Exercise 3 Compare these two ways of expressing the same idea:

In everyday speech or writing:	In science:
(1a) circles *which increase*	(1b) *increasing* circles
(2a) circles *which increase all the time*	(2b) *ever-increasing* circles

Rewrite these phrases in the same way as shown in example (b):

1 frequency which increases
2 waves which increase all the time
3 circles which decrease all the time
4 a vibration which increases slowly
5 a frequency which changes all the time
6 a number which grows all the time
7 molecules which vibrate
8 a velocity which increases rapidly
9 sounds which gradually decrease
10 resistance which gradually grows
11 a point which is always moving
12 a distance which is rapidly decreasing

13 the flow which is rapidly increasing
14 a seed which is slowly developing
15 a rate which is always changing

Exercise 4 Fill in the blank in each sentence with ONE suitable word:

1 Sound cannot exist without —— .
2 Sound travels through a —— .
3 Sound is made by —— and travels in —— .
4 When sound reaches the ear, the nerves send —— to the brain which —— them as sound.
5 No sound can be heard if the source of sound vibrates in a —— .
6 The —— of sound is affected by the medium through which it travels.
7 Sound travels —— in air than it does in a solid.
8 Lightning and thunder occur —— but we hear thunder later.
9 The velocity of sound is approximately 330 metres —— second.
10 The velocity of light is considerably —— than this.

Exercise 5 Answer these questions without referring to the Texts:

1 By what is sound made?
2 In which direction do sound waves travel from the source?
3 What happens when the waves reach the ear?
4 Why can no sound be heard in a vacuum?
5 What do we call the molecules of air, water or solid in which sound waves travel?
6 By what is the velocity of sound affected?
7 Explain why lightning at a distance can be seen before the thunder can be heard.
8 What is the approximate velocity of sound in air?
9 Would sound travel faster or slower than this in water?

Exercise 6 Questions for further discussion:

1 If a tree fell in a forest, and there was no living thing near enough to hear it, would it make any noise in falling?
2 How do you explain an echo? Give examples.
3 Does the length of a string on a guitar affect the note?
4 Which gives a higher note, a short pipe or a long one?
5 Why must astronauts on the moon talk to each other by radio instead of shouting?
6 Why do men working with jet aircraft at airports wear helmets covering their ears?

7 Do fish have ears? How can they hear in water?
8 How is the depth of the sea measured?
9 Why are ships' sirens blown at night in the Arctic?
10 Why does your voice echo in an empty hall, room or theatre, but not when it is full of furniture or people?
11 How could you calculate approximately how far away you are from the centre of a thunderstorm?
12 What materials could be used in buildings to deaden sound?

Vocabulary

to deaden echo helmet siren

Exercise 7 Suggestions for further activities:

1 Stand in the corner of a room facing the walls closely. Speak or sing and notice the difference in the sound of your voice.
2 Take a fairly large square of cardboard and ask someone to hold it in front of your face about 20cm away. Close your eyes and sing one long note. He will remove the cardboard and replace it, as you hold the note. Can you hear any difference in the sound of your voice? Can you tell without looking whether the cardboard is in front of your face or not? Explain why.
3 Take two bottles of the same size and half-fill one with water. Blow hard across the neck of each bottle and listen to the notes produced. Which is the higher note? Try this with empty bottles of different sizes, or with bottles with different quantities of water in them. Which produce the higher and lower notes?

Revision Exercises VI (Units 23–28)

I Give the meaning in your own language of these words:

1 ear-drum
2 characteristic
3 thunder
4 to multiply
5 nerve
6 sound
7 sphere
8 lightning
9 direction
10 tissue
11 upside down
12 brain
13 parallel
14 nitrogen
15 to collect
16 relative
17 force
18 ammonia

II Explain the meaning of:

1 the nutrition of plants
2 air resistance
3 an object of high density
4 maximum velocity
5 frequency of a wave
6 the force of gravity
7 natural phenomena
8 it merely rises and falls
9 regardless of the duration of the fall
10 a magnifying lens
11 the focal point of a lens
12 an inverted image
13 a spherical mirror
14 it transmits impulses
15 the image is laterally inverted
16 if it were not for bacteria
17 V = FL

III Give ONE word meaning:

1 living tissue
2 there ready to use
3 slightly changes
4 (which is) exactly the same
5 place where something comes from
6 getting bigger all the time
7 speeding up
8 material which carries sound
9 bend towards each other
10 taken in and used
11 bend away from each other
12 give a meaning
13 used up and finished
14 do a sum to find out
15 number of waves per second
16 happenings (in nature)

IV Answer these questions without referring to the Texts:

1 What does the greater part of the atmosphere consist of?
2 What is the approximate rate of acceleration of a freely-falling body?
3 What will modify this rate, at high velocities?
4 Only when can nitrogen be assimilated by animals and plants?
5 What shape is a convex lens?

6 What shape is a magnifying lens?
7 How might you burn a hole in a piece of paper using a convex lens?
8 What do parallel light rays do when passing through a concave lens?
9 What do parallel light rays do when reflected by a concave mirror?
10 If you stand in front of a plane mirror, would your image be (a) identical? (b) inverted? (c) magnified? (d) laterally inverted?
11 Which three characteristics have all waves in common?
12 How could you calculate the velocity of any wave?
13 What is meant by the term 'high-frequency wave'?
14 How do we hear a sound?
15 Why do we hear the thunder after we see the lightning?
16 Approximately what is the velocity of sound in air? Is it more or less than this in water?
17 Why is there no sound in a vacuum?
18 What do plants use combined nitrogen for?

V Choose the correct word to complete each of these sentences:

1 An unsupported body falls towards the centre of the earth because it is —— by the force of gravity.
(a) resisted (b) modified (c) attracted (d) forced
2 At high velocities, the resistance of the air —— the acceleration of a freely-falling body.
(a) attracts (b) produces (c) causes (d) modifies
3 The nitrogen required by plants and animals to build up their tissues would not be —— in sufficient quantities.
(a) available (b) able (c) remained (d) exhausted
4 When certain bacteria in the soil die, the nitrogenous compounds remain in the soil for the —— of plants.
(a) elements (b) nutrition (c) quantities (d) atmosphere
5 The spot where the sun's rays are —— by a convex lens is known as its focal point.
(a) concentrated (b) reflected (c) diverged (d) inverted
6 Focal length —— with the thickness of the lens.
(a) modifies (b) strikes (c) differs (d) collects
7 A mirror must be —— if it is to reflect an identical image.
(a) parallel (b) spherical (c) lateral (d) plane
8 The velocity of sound is —— by the medium through which it travels.
(a) vibrated (b) affected (c) interpreted (d) disturbed
9 Radio waves possess a —— frequency and travel at the velocity of light.
(a) high (b) low (c) dense (d) circular

10 The velocity of any wave can be —— by multiplying its frequency by its length.
(a) modified (b) affected (c) calculated (d) increased

11 A sound cannot exist without —— .
(a) a current (b) a vacuum (c) movement (d) gravity

12 Sound travels —— in air than in a liquid or a solid.
(a) faster (b) higher (c) lower (d) slower

13 Nitrogen cannot be —— by plants and animals unless it is combined with other chemical elements.
(a) required (b) assimilated (c) composed (d) exhausted

14 A plane mirror will reflect an —— but laterally inverted image.
(a) important (b) increasing (c) approximate (d) identical

15 Sound cannot travel in a —— .
(a) medium (b) fluid (c) solid (d) vacuum

16 The greater part of the atmosphere consists of —— .
(a) oxygen (b) nitrogen (c) ammonia (d) hydrogen

17 If a person is short-sighted, he cannot see —— objects clearly
(a) near (b) inverted (c) magnified (d) distant

18 The distance between the crest and the lowest point of a wave is known as its —— .
(a) amplitude (b) frequency (c) reflection (d) density

VI Use a more scientific term in place of each word or phrase italicised in these sentences:

1 A *flat* mirror reflects an identical but laterally inverted *picture* of the *thing* in front of it.

2 Sound is *carried* to the ear and nerves *send messages* to the brain, which *gives them a meaning* as sound.

3 Light waves which *hit* a mirror are *thrown back.*

4 *Really,* there is a *greatest speed* at which an object will fall in air, *no matter for how long it falls.*

5 The *end speed* depends on the shape and relative *compactness* of the falling body.

6 The greater part of the *air around the earth is made* of nitrogen.

7 *Some* bacteria on the roots of *some kinds* of plants fix free nitrogen and *join it together* chemically with other elements.

8 *A very large number of radio waves come in one second.*

9 The *speed* of sound is *much* less than that of light; it is *about* 330 metres *in every* second.

10 When a body is *not held up,* it falls towards the centre of the earth because it is *pulled* by the force of gravity.

Unit 29 Magnetism

Vocabulary

a bar | iron filings | to sprinkle | steel
to behave

A

When a bar of iron *pulls towards itself* other small pieces of iron, it *behaves like* a magnet, and this *pull* is called magnetism. The main force of the pull *is found* at the two ends of the bar, which are known as the north and south poles. The area in which the pull works is known as the magnetic field.

The field of magnetic force can be *shown* by *putting* a piece of glass over a magnet and sprinkling iron filings over the surface of the glass. The curved lines formed by the iron filings *show where* the magnetic lines of force *are*.

Magnetism which is *produced without touching* by an electric current is known as electro-magnetism. When a current of electricity is passed through a conductor, a magnetic field is formed around it. A piece of iron or steel which is put in this field *becomes a magnet* and is known as an electro-magnet.

In all magnets, the north pole *pushes away* all other north poles, and pulls towards itself all other south poles (and *this is true if said the other way round*). In other words, poles *which are the same* push each other away, and poles *which are not the same* pull towards each other.

Diagram of electro-magnet

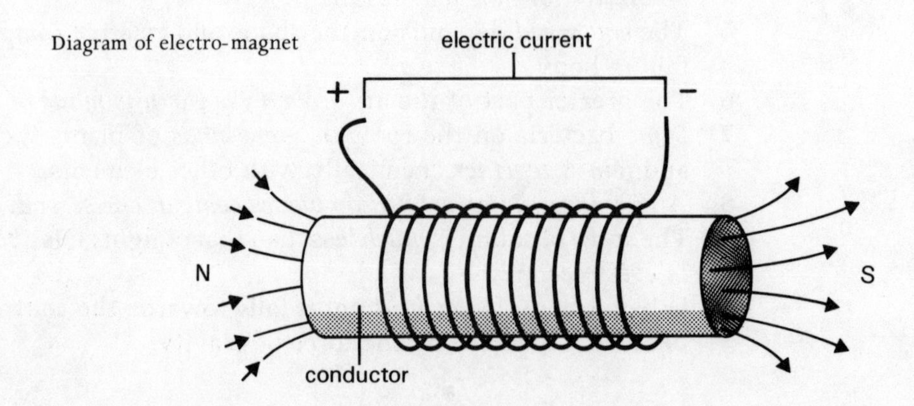

B

When a bar of iron (1)attracts other small pieces of iron, it (2)acts as a magnet, and this (3)attraction is called magnetism. The main force of the (3)attraction (4)lies at the two ends of the bar, which are known as the north and south poles. The area in which the (3)attraction works is known
5 as the magnetic field.

The field of magnetic force can be (5)demonstrated by (6)placing a piece of glass over a magnet and sprinkling iron filings over the surface of the glass. The curved lines formed by the iron filings (7)indicate the location of the magnetic lines of force.

10 Magnetism which is (8)induced by an electric current is known as electro-magnetism. When a current of electricity is passed through a conductor, a magnetic field is formed around it. A piece of iron or steel (6)placed in this field (9)is magnetised and is known as an electro-magnet.

In all magnets, the north pole (10)repels all other north poles, and
15 (1)attracts all other south poles (and (11)vice versa). In other words, (12)like poles (10)repel each other, and (13)unlike poles (1)attract.

Exercise 1 Find the way in which the words and phrases italicised in Text A are expressed in Text B:

1 pulls towards itself
2 behaves like
3 pull
4 is found
5 shown
6 putting
7 show where . . . are
8 produced without touching
9 becomes a magnet
10 pushes away
11 this is true if said the other way round
12 which are the same
13 which are not the same

Exercise 2 AND VICE VERSA (Latin) is often used in scientific writing at the end of a sentence, meaning that, if the sentence were expressed the other way round, it would still be true:

(a) South poles repel south poles and attract north poles, *and vice versa*.
Here, AND VICE VERSA means:
(b) North poles repel north poles and attract south poles.

That is, we are saying the same thing, but the other way round. Now express fully the meaning of AND VICE VERSA in these sentences:

1 A dull surface absorbs more heat and reflects less heat than a shiny one, and vice versa.
2 The moon reflects sunlight to the earth, and vice versa.
3 A good heat insulator is a poor conductor of heat, and vice versa.
4 A poor heat insulator is a good conductor of heat, and vice versa.
5 In a vacuum flask, the heat of the interior cannot reach the exterior, and vice versa.
6 The image of an object in a plane mirror is identical, but the left side is on the right, and vice versa.
7 Short waves have high frequencies, and vice versa.
8 Long waves have lower frequencies, and vice versa.

More nouns ending with -ion

Here are some more verbs which form nouns ending in -ION:

a

VERB	NOUN
(i) Drop E, add -ION:	
to concentrate	concentrat-ion
to locate	locat-ion
to calculate	calculat-ion
to illustrate	illustrat-ion
to vibrate	vibrat-ion
to indicate	indicat-ion
to demonstrate	demonstrat-ion
(ii) Drop E, add -TION:	
to induce	induc-tion
(iii) T to S/SS, add -ION:	
to transmit	transmiss-ion
to invert	invers-ion
to convert	convers-ion

b

VERB	NOUN
(i) Add -ATION:	
to fix	fix-ation
to form	form-ation
to interpret	interpret-ation
(ii) Drop E, add -ATION:	
to condense	condens-ation
to derive	deriv-ation
to preserve	preserv-ation
to combine	combin-ation
to magnetise	magnetis-ation
(iii) Y to I, add -CATION:	
to magnify	magnifi-cation
to modify	modifi-cation

Exercise 3 Each of these nouns can be used, once only, to complete each of these sentences:

1 The curved lines formed by iron-filings over the magnetic field indicate the —— of the lines of force.
2 Certain bacteria are of great importance for the —— of atmospheric nitrogen.
3 Radio is a good example of the —— of electro-magnetic waves into sound waves.
4 Dipping a wooden spoon into boiling water and seeing how the handle remains cool is a good —— of the fact that wood is a poor heat conductor.
5 The —— of an iron bar can be induced by placing it in a magnetic field.
6 A piece of iron magnetised in this way is said to have been magnetised by —— .
7 The heat from the —— of the sun's rays passing through a convex lens can burn a hole in a piece of paper.
8 A vacuum flask may be used for the —— of heat or cold.
9 Hearing a sound is the result of the —— of the ear-drum, the —— of impulses by the nerves, and the —— of these impulses by the brain.
10 The formula for the —— of the velocity of any wave is V = FL.
11 The curved lines formed by iron-filings near a magnet are an —— of the lines of magnetic force in the field.
12 The —— of an image may be demonstrated by holding a white card at the focal point of a convex lens.
13 A —— of wave motion can be given by dropping a stone into a smooth lake.
14 Heat from an open fire is transferred to a room by a —— of all three methods of heat transfer.
15 Water-vapour returns to liquid water in a process known as —— .
16 Heating a fluid will result in the —— of convection currents.
17 The —— in the acceleration of a free-falling body in air will depend on its shape and relative density.
18 The strength of —— of a convex lens depends on its thickness.
19 The table opposite shows the —— of nouns from their verbs.

Exercise 4 The four verbs given below can all replace the verb TO SHOW, but there is a difference in meaning between them:

TO INDICATE		to show by pointing
TO DEMONSTRATE	means	to show how something works, happens or behaves
TO PROVE		to show some fact to be true
TO ILLUSTRATE		to show with a picture or example

Use the correct form of the most suitable of these verbs to replace the SHOW phrase in each of these sentences:

1 We can show it to be true that heat rays from the sun are not themselves hot, because we know that space remains cold although the rays are passing through it.
2 We can show by example that silver is a good conductor of heat, by dipping a silver spoon into boiling water and noticing that the handle rapidly becomes hot.
3 We can show how *the inversion* of an image happens by holding a white card at the focal point of a convex lens.
(Rewrite this sentence using a passive form.)
4 How can the location of the lines of magnetic force be shown?
5 Dropping a stone into a smooth lake is a good way of showing waves rising and falling.
6 Dropping a stone from a height is a good way of showing an example of the attraction of the force of gravity on a falling body.
7 How can we show it to be true that light travels much faster than sound?
8 Thunder and lightning show by example that light travels much faster than sound.
9 We can show how *the force* of gravity works by throwing a ball into the air.
(Rewrite this sentence in the passive.)
10 The fact that no sound is produced in a vacuum shows it to be true that sound waves cannot travel without a medium.
11 The arrows in the diagram of a wave showing by pointing the amplitude and length of a wave.
12 The fact that water expands on heating can be shown to be true by experiment.

Exercise 5 Rewrite this summary, using passive forms, and also putting more scientific terms in place of the words and phrases italicised:

When an iron bar *pulls towards itself* other pieces of iron, it *behaves like* a magnet. We call this *pull* magnetism. We find its main force around the two *ends* of the magnet, and we call the magnetised area the magnetic field. We can *show where* the lines of force *are* by sprinkling iron filings on a piece of glass which we have *put* over the magnet.

We can *produce* electro-magnetism *without touching*, by passing an electric current through a conductor, and we can magnetise a piece of iron by *putting* it in this *area of force*.

A like pole *pushes away* a like pole and *pulls towards itself* an unlike pole.

Exercise 6 Answer these questions without referring to the Texts:

1 What is magnetism?
2 Where does the main force of magnetic attraction lie?
3 Where are the north and south poles of a magnet?
4 What do we call the area in which the forces of attraction work?
5 How can the location of the magnetic lines of force be demonstrated?
6 What is electro-magnetism?
7 How can magnetism be induced in a piece of iron?
8 If two south poles were placed near each other, what would happen?
9 If a north and a south pole were placed near each other, what would happen?

Exercise 7 Questions for further discussion:

1 Have you ever seen any toys or games which use magnets? If you have, describe one. Can you explain simply how it works?
2 What do you know about a compass, how it works, and what it is used for?
3 How do you use a compass to find your way when you are lost?
4 How could you use a compass to find out which is the south pole of any magnet?

Vocabulary

compass

Exercise 8 Suggestions for further activities:

1 Make a simple compass by stroking a large needle with a magnet – in one direction only. Hang the needle horizontally from a thread tied round the middle. Wait for the needle to come to rest. It will then point in a north–south direction.
2 What happens if you put a piece of iron or steel (such as a knife) near to a compass needle pointing north?
3 Place a small compass in a glass dish and see whether a magnet under it will attract the needle. Repeat the experiment using an iron dish in place of the glass dish.

Unit 30 Movement of the Earth

Vocabulary

angle	the east	planet	tilted
axis	path	to set (sun)	the west
axes (pl.)			

A

If you are travelling forwards in a train, *the things around you* outside *seem* to be moving backwards. But this is not *really true*. It is the train which is moving.

It is like this with the earth. The sun seems to rise in the east, move 5 across the sky, and set in the west. In other words, the sun seems to travel round the earth. However, this is not really true. It is the earth which is *turning* on its own axis. *The turning* of the earth on its own axis *is what makes* the change from day to night.

The earth also travels round the sun on an *unchanging* path, known as 10 an orbit. *As well as* the earth, there are eight other planets of different sizes, *which all turn round* on their own axes and *which all travel round* the sun in their own unchanging orbits. These are known as the solar system. The planet nearest the sun is Mercury, and *the planet farthest from the sun is* Pluto.

15 The orbits of the earth and the other planets are not perfect circles. *They are ovals.* The earth's axis is not *at an angle of* 90° to its orbit; *but it is tilted a little.* The turning of the earth round the sun at this angle and in an *oval-shaped* orbit is what makes the change of the seasons.

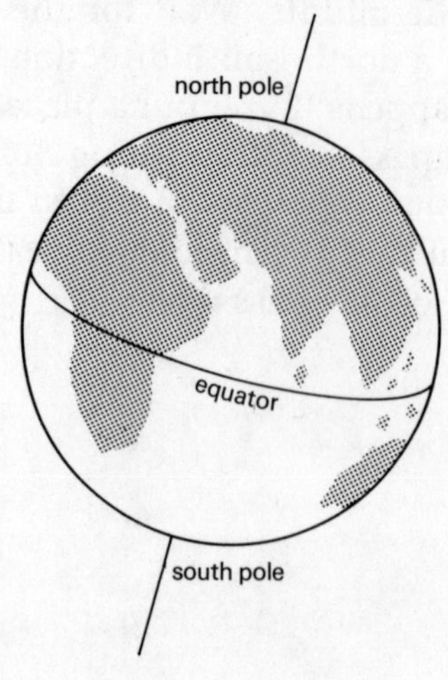

Diagram of the earth

B

If you are travelling forwards in a train, (1)surrounding objects outside (2)appear to be moving backwards. But this is not (3)in fact the case. It is the train which is moving.

(4)So it is with the earth. The sun (2)appears to rise in the east, move
5 across the sky, and set in the west. In other words, the sun (2)appears to travel round the earth. (3)In fact, however, this is not the case. It is the earth which is (5)rotating on its own axis. (6)The rotation of the earth on its own axis (7)causes the change from day to night.

The earth also travels round the sun on a (8)fixed path, known as an
10 orbit. (9)In addition to the earth, there are eight other planets of different sizes, (10) all rotating on their own axes and (11)round the sun in their own (8)fixed orbits, and which are known as the solar system. The planet nearest the sun is Mercury and the (12)most distant, Pluto.

The orbits of the earth and the other planets are not perfect circles,
15 (13)but ellipses. The earth's axis is not (14)at right-angles to its orbit (15)but slightly tilted. (6)The rotation of the earth around the sun at this angle and in an (16)elliptical orbit (7)causes the change of the seasons.

Exercise 1 Find the way in which the words and phrases italicised in Text A are expressed in Text B:

1 the things around you
2 seem
3 really true
4 It is like this
5 turning
6 the turning
7 is what makes
8 unchanging
9 As well as
10 which all turn round
11 which all travel round
12 planet farthest from the sun is
13 They are ovals
14 at an angle of 90°
15 but it is tilted a little
16 oval-shaped

Exercise 2 Two sentences can sometimes be joined like this:

(a) The orbit is not a perfect circle. (*But*) *It is* an ellipse.
(b) The orbit is not a perfect circle *but* an ellipse.

Join these pairs of sentences in the same way as in example (b):

1 The planets rotate not only on their own axes. But they also rotate around the sun.
2 There is not only the earth in the solar system. But there are also eight other planets.
3 The sun appears not only to move round the earth. But it also appears sometimes to be smaller than the moon.
4 The earth is rotating not only on an elliptical orbit. But it is also rotating on a tilted axis.
5 These facts not only cause the change from day to night. But they also cause the change of one season to another.

Exercise 3 Notice the way in which the relative clause italicised in (a), can be contracted in (b):

(a) Objects *which surround you* appear to be moving.
(b) *Surrounding* objects appear to be moving.

Rewrite these sentences, contracting the italicised clauses, as in example (b):

1 If you are in a train *which is moving*, objects *which surround you outside* appear to be moving.
2 The sun *which is rising* appears in the east, and the sun *which is setting* is seen in the west.
3 The earth *which is rotating* follows a fixed orbit around the sun.
4 This orbit *which is fixed* is not circular but elliptical.
5 The solar system contains nine planets *which are rotating*.
6 They all rotate in orbits *which do not change*.
7 The rotation of the earth on an axis *which is tilted* and in an orbit *which is elliptical* causes the change of the seasons.

Exercise 4 Read or rewrite this passage, using a more 'scientific' term for each word or phrase italicised:

When your train is moving forwards, outside *things seem* to move backwards; *but/really* this is not *true*. The earth *moves* round the sun and also *turns round* on its own axis. Its *path* round the sun is *oval-shaped* and *unchanging*, and its axis is not *at an angle of 90°* to its *path* but is tilted *a little*. These two facts *make* the changes both of light and dark and of winter and summer.

Exercise 5 Answer these questions without referring to the Texts:

1 When you are moving forwards in a train, are surrounding objects outside in fact moving backwards?
2 To what can we compare this?
3 What does the sun appear to do, and what is in fact the case?
4 What causes the change from day to night?
5 What do you understand by 'the earth's orbit'?
6 What is the name given to the sun and all its planets?
7 How many planets are there rotating around the sun?
8 What is the name of the planet (a) nearest and (b) most distant from the sun?
9 What shape are the orbits of the planets?
10 What causes the change of seasons on earth?

Exercise 6 Questions for further discussion:

1 How long does it take for the earth to rotate once on its own axis?
2 How long does it take the earth to rotate once round the sun?
3 Why do we have a Leap Year (i.e. a year in which February has 29 days) once every four years?
4 Why is it hotter at the equator than at the poles?
5 Approximately how far away is the sun from the earth?
6 When it is snowing in December in Canada, what is the weather like in Australia? Why?

Vocabulary
equator

Exercise 7 Suggestions for further activities:

1 Make a simple sun clock. Drive a staight stick about four feet long into a flat piece of sunlit ground, so that it is at right angles to the ground. Tie the end of a piece of string about 15–20 feet long to the base of the stick. About two or three hours before noon, measure with the string the length of the shadow of the stick. Using this as a radius, draw a half-circle on the ground. Mark the point where the shadow touches the half-circle. Repeat this in the afternoon. Find the point on the half-circle which is exactly half-way between these two points, and stretch the string between the stick and the half-way mark. This is a true north–south line. When the shadow of the stick falls onto this line, it is exactly twelve noon by the sun. Is it also twelve noon by your watch?
2 Draw a simple diagram of the solar system.

Unit 31 Atoms and Molecules

Vocabulary

a charge	negative	positive	protein
chemistry	neutral		

A

An element is a substance which cannot be *broken down* into other substances. Elements are *made up of* atoms, which consist mainly of *emptiness*.
In the middle of each atom there is a nucleus, in which are found mainly (a) protons, *which have* a positive electric charge, and (b) neutrons, which
5 have no charge. Around the nucleus are electrons, which rotate very rapidly and carry a negative electric charge.
The number of protons is *the same as the number* of electrons, and *therefore* the whole atom is neutral *in electricity*. *When we say that we mean* that the whole atom has no charge.
10 The atom of each element *is different* from *the atom* of every other element in the size and weight of its nucleus and the number of its electrons. The simplest atom is the atom of the element hydrogen, which has *only one* electron. Other elements have more *complicated* atoms with a larger nucleus and *more* electrons, e.g. uranium, which has 92.
15 Most substances are *made up of* two or more elements chemically combined as molecules *so that they form* compounds – the most *often found/is* water. A molecule of water is made up of 3 atoms – 2 of hydrogen and 1 of oxygen (*written like this in chemistry:* H_2O).
Some elements have molecules of more than one atom, e.g. oxygen *of*
20 *the atmosphere/is* in pairs of atoms (written like this in chemistry: O_2). Some compounds, such as proteins, have hundreds of atoms of several elements in each of their molecules.

B

An element is a substance which cannot be (1)decomposed into other substances. Elements are (2)composed of atoms, which consist mainly of (3)space.

(4)At the centre of each atom there is the nucleus, in which are found mainly (a)protons, (5)possessing a positive electric charge, and (b)neutrons, (5)possessing no charge. Around the nucleus are electrons, which rotate very rapidly and carry a negative electric charge.

The number of protons is (6)equal to that of electrons and (7)consequently the whole atom is (8)electrically neutral. (9)By this is meant that the whole atom has no charge.

The atom of each element (10)varies from (11)that of every other element in the size and weight of its nucleus and the number of its electrons. The simplest atom is (11)that of the element hydrogen, which (5)possesses (12)a single electron. Other elements (5)possess more (13) complex atoms with a larger nucleus and (14)a greater number of electrons, e.g. uranium, which has 92.

Most substances are (2)composed of two or more elements chemically combined as molecules (15)to form compounds – the most (16)common (17) being water. A molecule of water is (2)composed of 3 atoms – 2 of hydrogen and 1 of oxygen ((18)chemical formula: H_2O).

Some elements (5)possess molecules of more than (12)a single atom, e.g. (19)atmospheric oxygen (20)exists in pairs of atoms ((18)chemical formula: O_2). Some compounds, such as proteins, (5)possess hundreds of atoms of several elements in each of their molecules.

Exercise 1 Find the way in which the words and phrases italicised in Text A are expressed in Text B:

1 broken down
2 made up of
3 emptiness
4 In the middle
5 which have
6 the same as the number
7 therefore
8 in electricity
9 When we say that we mean
10 is different
11 the atom
12 only one
13 complicated
14 more
15 so that they form
16 often found
17 is
18 written like this in chemistry
19 of the atmosphere
20 is

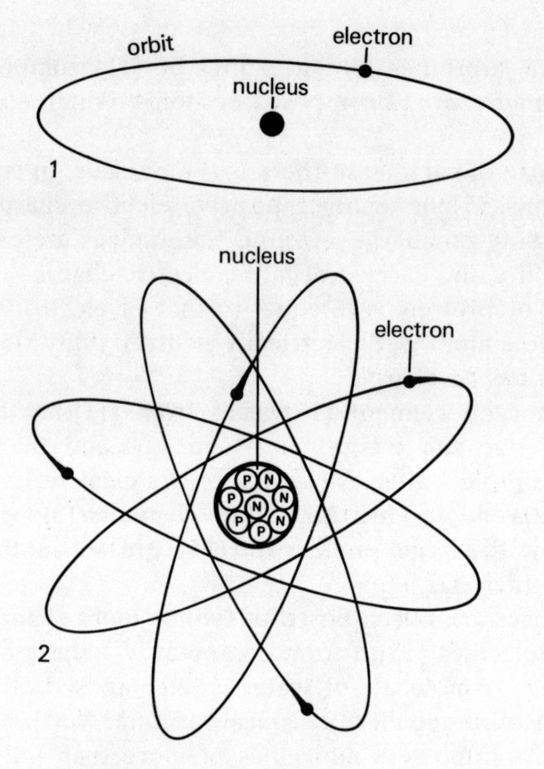

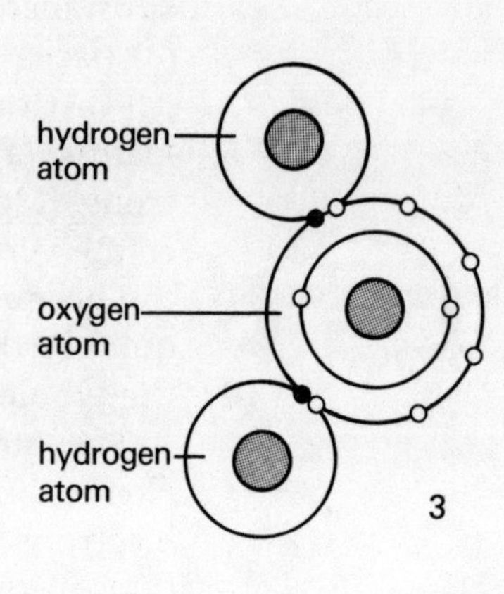

1. Diagram of hydrogen atom
2. Diagram of atom with four electrons
3. Water molecule

Exercise 2 POSSESS is often used to replace the verb 'have':

(a) Protons *have* a positive electric charge.
(b) Protons *possess* a positive electric charge.

Read or rewrite these sentences, using the correct form of POSSESS in place of 'has/have':

1 Neutrons do not have an electric charge.
2 Since atoms have the same number of protons as electrons, the whole atom is electrically neutral.
3 If the element hydrogen has a single electron, this means that it must have a single proton.
4 Other elements have more complex atoms which have a larger number of protons and electrons.
5 A complex atom has a larger nucleus.
6 Uranium is an example of an element which has a large number of electrons.

7 Atmospheric oxygen has a molecule of two atoms.
8 All waves have three characteristics in common.

Exercise 3 THAT is often used to avoid repeating a singular noun when it occurs more than once in the same sentence:

(a) The number of electrons equals *the number* of protons.
(b) The number of electrons equals *that* of protons.

Read and rewrite these sentences, using THAT to avoid repeating the italicised nouns:

1 The simplest *atom* is the atom of hydrogen.
2 A *molecule* of water is smaller than the molecule of a protein.
3 A *molecule* of water possesses 3 atoms, and the molecule of carbon dioxide also possesses 3.
4 The *electric charge* of a proton is positive, but the electric charge of an electron is negative.
5 The *orbit* of an electron may be compared to the orbit of a planet.
6 The *planet* of Mercury is much nearer to the sun than the planet of Jupiter.
7 The *velocity* of light is much greater than the velocity of sound.
8 The *frequency* of light waves is much higher than the frequency of sound waves.
9 The *velocity* of electro-magnetic waves is equal to the velocity of light.
10 The *reflection* of an object in a concave mirror is different from the reflection in a plane mirror.
11 The *focal length* of a thick convex lens is shorter than the focal length of a thin one.
12 The *acceleration* of a free-falling body of high density is greater than the acceleration of a free-falling object of low density.

Exercise 4 Rewrite this summary, using passive forms (with agent only where necessary). (The subjects of the passive sentences are italicised):

You cannot decompose *an element* into other substances. Elements consist of atoms, at the centre of which we find *the nucleus.* We find *the proton and the neutron* inside the nucleus. The proton carries *a positive charge,* and the electron, which rotates very rapidly round the nucleus, carries *a negative charge.* Hundreds of atoms of several elements form *compounds such as protein.* Two atoms of hydrogen and one of oxygen form *one molecule of water.*

Exercise 5 KNOWN AS . . .: Can you complete each sentence with ONE appropriate word?

1 A substance which cannot be decomposed into other substances is known as an —— .
2 The tiny particles of which elements are composed are known as —— .
3 The centre of an atom is known as the —— .
4 The part of the atom which carries a positive charge is known as a —— .
5 The part of the atom which has neither a positive nor a negative charge is known as a —— .
6 The rotating part of the atom which carries a negative charge is known as an —— .
7 Atoms of two or more elements which are chemically combined are known as —— , and these form substances known as —— .

Adjectives ending with -ar

These words may make an adjective ending with -AR:

NOUN	ADJECTIVE
circle	circ-ular
nucleus	nucle-ar
molecule	molecul-ar

Notice the change in pronunciation, thus: the stress moves to the second syllable MOL-ec-ule – mol-EC-ular.

Exercise 6 Read and rewrite these sentences, using one of the above adjectives to replace each phrase italicised:

1 Most chemistry books contain a table showing weights *of molecules.*
2 Energy *of the nucleus* can be converted to heat.
3 The orbit of the earth is not *shaped in a circle* but is an ellipse.
4 Vibration *of molecules* can be accelerated by heating.

Exercise 7 Answer these questions without referring to the Texts:

1 What is (a) an element? (b) a molecule? (c) a compound?
2 Which is the commonest compound?
3 Of what does the nucleus of an atom mainly consist?
4 Which part of an atom is charged (a) positively? (b) negatively?
5 Why is the whole atom electrically neutral?
6 In what do the atoms of different elements vary?
7 How many electrons has an atom of hydrogen?
8 Why is the chemical formula for water H_2O?
9 Why is the chemical formula for atmospheric oxygen O_2 instead of simply O?
10 How many electrons has an atom of uranium?

Exercise 8 Questions for further discussion:

1 Approximately how many elements do you think make up all the substances on earth?

2 (a) What do you think is the commonest element?
(b) How many elements do you think are fairly common?

3 Which of these substances are elements and which are compounds?

(a) common salt
(b) carbon dioxide
(c) combined nitrogen
(d) water
(e) oxygen
(f) hydrogen
(g) protein
(h) nitrogen
(i) ammonia
(j) potassium permanganate

4 If all the empty space in all the atoms of all the substances on earth were removed, how big do you think the remaining substance would be?

5 Are all elements natural?

Exercise 9 Suggestions for further activities:

Compare your diagram of the solar system with the diagrams of the atoms in Figure 20.

Unit 32 Static Electricity

Vocabulary
to mention

A

When a plastic ruler has been rubbed against wool, it will attract small pieces of paper. To explain why this *occurs,* we must *look back* to our knowledge of the balance *of electricity* of the atom.

We have explained before that electrons *travel round* the nucleus, just
5 as planets travel round the sun. But there is a difference: *the planets* (*mentioned second*)/*keep* their orbits by the attraction *of gravity,/but unlike this/the electrons* (*mentioned first*) keep their orbits by the attraction of electricity, since unlike charges attract each other (*that is,* the negative charge of the electron is attracted by the positive charge of the proton),
10 *in this way* making the whole atom electrically neutral.

But if electrons are *taken away* from, or added to, an atom, it will then carry an electric charge, and charged atoms *of this kind* are known as ions.

The simplest method of *doing* this is by *rubbing.* Electrons are *knocked out of* the atoms of the plastic ruler, leaving them with too few, and in this
15 way carrying a positive charge. Electrons are added to the atoms of the wool, in this way giving them a negative charge.

In this way we have *got* an electric charge. An electric charge *made* in this way is known as static electricity.

B

When a plastic ruler has been rubbed against wool, it will attract small pieces of paper. To explain why this (1)takes place, we must (2)refer to out knowledge of the (3)electrical balance of the atom.

(4)As explained previously, electrons (5)orbit the nucleus, just as
5 planets (5)orbit the sun. But there is a difference: (6)the latter (7)maintain their orbits by (8)gravitational attraction, (9)whereas (10)the former (7)maintain their orbits by (3)electrical attraction, since unlike charges attract each other ((11)i.e. the negative charge of the electron is attracted by the positive charge of the proton), (12)thus making the whole atom
10 electrically neutral.

But if electrons are (13)removed from, or added to, an atom, it will then carry an electric charge, and (14)such charged atoms are known as ions.

The simplest method of (15)achieving this is by (16)friction. Electrons are (17)dislodged from the atoms of the plastic ruler, leaving them with
15 too few, and (12)thus carrying a positive charge. Electrons are added to the atoms of the wool, (12)thus giving them a negative charge.

(12)In this way an electric charge has been (18)obtained. An electric charge (12)thus (19)produced is known as static electricity.

Exercise 1 Find the way in which the words and phrases italicised in Text A are expressed in Text B:

1 occurs
2 look back
3 of electricity
4 We have explained before that
5 travel round
6 the planets (mentioned second)
7 keep
8 of gravity
9 but unlike this
10 the electrons (mentioned first)
11 that is
12 in this way
13 taken away
14 of this kind
15 doing
16 rubbing
17 knocked out of
18 got
19 made

Exercise 2 A sentence may be contracted in order to avoid repeating a noun:

(a) Electrons may be *removed from an atom or added to an atom.*
(b) Electrons may be *removed from, or added to,* an atom.

Read and rewrite these sentences, contracting them to avoid repeating the nouns italicised, as in example (b):

1 Heat may be radiated by *a solid* or conducted by *a solid.*
2 Very little heat can penetrate *a vacuum flask* or escape from *a vacuum flask.*
3 Iron filings will be attracted to *a magnet* and can be picked up by *a magnet.*
4 An electron may be dislodged from *an atom* or added to *an atom* by friction.
5 Electrons are equal in number to *protons* and attracted by *protons.*

THE FORMER/THE LATTER

In order to avoid repetition when referring to two things mentioned in a sentence, THE FORMER is used to replace the thing mentioned first, and THE LATTER is used to replace the thing mentioned second:

(a) *Electrons* orbit the nucleus just as *planets* orbit the sun.
(b) *The former* (i.e. electrons) rotate an elliptical orbit and so do *the latter* (i.e. planets).

Exercise 3 Read and rewrite these sentences, avoiding the repetition of the words italicised by using THE FORMER/THE LATTER, as in example (b):

1 When electrons are dislodged from the atoms of the plastic ruler and added to those of the wool, *the atoms of the plastic ruler* now carry a positive charge while *the atoms of the wool* carry a negative one.
2 In any magnet there is a south pole and a north pole. *The north pole* repels all other north poles and *the south pole* repels all other south poles.
3 Two of the planets in the solar system are called Mercury and Pluto. *Pluto* is the farthest from, and *Mercury* is the nearest to, the sun.
4 An element is composed of similar atoms, and a compound is composed of different atoms. *An element* cannot be decomposed into other substances, but *a compound* can.
5 An atom of hydrogen possesses a single proton, whereas an atom of uranium possesses 92. Therefore, *an atom of uranium* is much heavier than *an atom of hydrogen.*
6 A football will fall more slowly than a stone of the same size, because *the football* has a lower density than *the stone.*

7 A convex lens is thicker in the middle than at the edges; a concave lens is thinner in the middle than at the edges. *The convex lens* will cause parallel light rays passing through it to converge, and *the concave lens* will cause them to diverge.

8 Thunder and lightning occur simultaneously, but at a distance *lightning* can be seen before *the thunder* can be heard.

9 The nucleus of an atom is mainly composed of protons and neutrons. *Protons* carry a positive charge, while *neutrons* have no electric charge.

10 Both planets and electrons rotate in elliptical orbits, but *planets* maintain their orbits by gravitational, and *electrons* maintain theirs by electrical, attraction.

Exercise 4 Answer these questions without referring to the Texts:

1 By what force do electrons maintain their orbits round the nucleus?
2 In what way do the planets differ from this in their orbits round the sun?
3 Why is the electron attracted to the proton?
4 When a plastic ruler is rubbed on wool, why do the atoms of the plastic ruler become positively charged?
5 Why do the atoms of the wool become charged negatively?
6 If an atom becomes negatively or positively charged, what is it then known as?
7 What is the main difference between an atom and an ion?
8 What is meant by the term 'static electricity'?

Exercise 5 Rewrite this summary of the Texts, using more 'scientific' terms in place of the words and phrases italicised. One small *change in word order is needed):

In order to explain how a charge of static electricity may be *got* by *rubbing*, we must *look back* to the electrical balance of the atom, which is *kept* because the negative charges of electrons *make neutral* the positive charges of the protons, *because* charges **which are not the same/pull towards each other*.

However, when electrons are *knocked out* from some atoms, by *rubbing*, and added to others, *electrically charged atoms* are formed, *in this way/making* a charge of static electricity.

Exercise 6 Questions for further discussion:

1 What causes lightning?
2 Why do clothes of synthetic fibres sometimes stick to you?
3 Find out about amber. Why is it of special importance in electricity?

Vocabulary
amber
fibre

Exercise 7 Suggestions for further activities:

1 Try the following experiments. Observe the results and try to explain them:
(a) Comb your hair (when it is newly washed) with a plastic comb.
(b) Rub a piece of plastic (e.g. a gramophone record) on a woollen jacket or silk dress. Hold it over some small pieces of paper.
(c) When undressing at night, turn out the light, and when your eyes are used to the darkness, take off the nylon/terylene/synthetic fibre clothes you have worn during the day. What can you see or hear?
Think of, or observe, other examples of static electricity.
2 Blow up a rubber balloon, and after tying up its end, rub it for a short time with a piece of silk or wool cloth. It should now stick to the wall.
3 Turn on a tap so that a thin stream of water flows from it. Take a comb and rub it on your hair or on a piece of wool or silk. Hold the comb 2–3cm away from the stream of water. The water will be attracted to the comb. Why?
4 Make a simple newspaper electroscope. Cut a strip of newspaper about 50cm long and 3cm wide. Fold it in half and put it on a table. Rub it hard with a piece of wool cloth or fur, and then hang it over a ruler. Bring a negatively charged rubber or plastic comb near to the end of the strip. If the paper carries a negative charge the ends will move away, if they are carrying a positive charge, they will move towards the rubber or comb. Why?

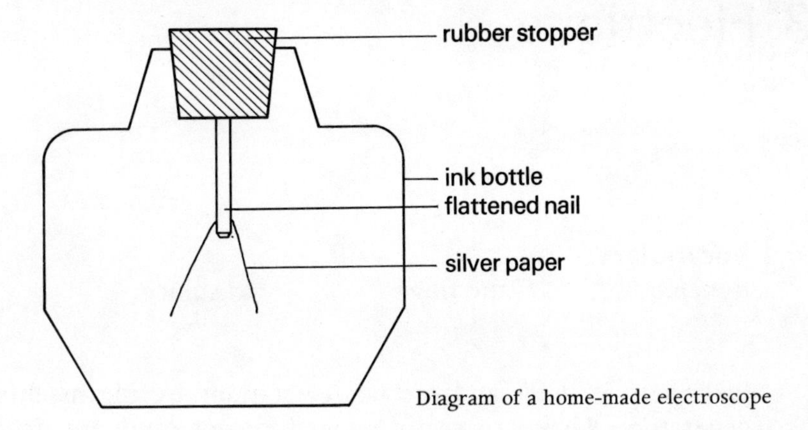

Diagram of a home-made electroscope

5 Make a real electroscope. Take a clean, empty ink bottle and a rubber stopper which just fits it. Cut two strips of silver paper about 3cm long and 5 or 6mm wide. Flatten a fairly large nail with a hammer and insert it into the stopper. Put a little nail varnish on each side of the end of the nail. Before it dries, pick up the pieces of silver paper by touching them with the varnished nail, so that one piece is stuck to each side. Place the stopper in the ink bottle, and the electroscope is ready. (See diagram).

Bring various materials near to the stopper and observe the movement of the strips of silver paper. They will diverge when a negatively charged body is brought near.

Vocabulary
nail varnish

Unit 33 Electricity

Vocabulary

dynamo the flow resistance

A

Electricity is *made* at large power stations by big machines known as generators. They are *really* large dynamos, and are driven by *strong machines* which *get* their power from water or steam. The electricity which gets its power from water is known as hydro-electricity.

5 The flow of electricity along a wire may be *said to be like* the flow of water along a pipe, and consequently it is known as current.

If you *think of* water *which flows* along a pipe, you may say that the volume of water *which passes* a *known* point in a *measured* time is *like* the electric current. We measure electric current in amperes.

10 The pressure of the water in the pipe may be said to be like the electric potential. We measure potential in volts.

The resistance of the walls of the pipe to the water current may be said to be like the resistance of the wire to the electric current; i.e. a narrow pipe *gives* more resistance than a wide pipe, and *in the same way*, a thin 15 wire gives more resistance than a thick wire of the same metal. We measure resistance in ohms.

B

Electricity is (1)generated at large power stations by big machines known as generators. They are, (2)in fact, large dynamos, and are driven by (3)powerful engines which (4)derive their power from water or steam. The electricity which (4)derives its power from water is known as hydro-
5 electricity.

The flow of electricity along a wire may be (5)compared to the flow of water along a pipe, and consequently it is known as current.

If you (6)consider water (7)flowing along a pipe, the volume of water (8)passing a (9)certain point in a (10)given time is (11)similar to the
10 electric current. Electric current is measured in amperes.

The pressure of the water in the pipe may be (5)compared to the electric potential. Potential is measured in volts.

The resistance of the walls of the pipe to the water current may be (5)compared to the resistance of the wire to the electric current; i.e. a
15 narrow pipe (12)offers more resistance than a wide pipe, and (13)similarly, a thin wire (12)offers more resistance than a thick wire of the same metal. Resistance is measured in ohms.

Exercise 1 Find the way in which the words and phrases italicised in Text A are expressed in Text B:

1 made
2 really
3 strong machines
4 get
5 said to be like
6 think of
7 which flows
8 which passes
9 known
10 measured
11 like
12 gives
13 in the same way

Exercise 2 Complete these sentences:

1 Temperature is measured in —— .
2 In science, temperature is always measured in —— .
3 Resistance is measured in —— .
4 Electric potential is measured in —— .
5 Electric current is measured in —— .
6 Wave-length is the distance between ——.
7 Frequency is the number of ——.
8 Wave amplitude is the distance between ——.
9 Wave velocity is calculated by multiplying ——.
10 The volume of a cube is calculated by multiplying ——.
11 The area of a square is calculated by multiplying ——.
12 The focal length of a lens is the distance between ——.
13 The acceleration of a free-falling body is approximately ——.

Exercise 3 Contract the WHICH + verb forms italicised in these sentences, as in example (b):

(a) Big machines *which generate* electricity are known as dynamos.
(b) Big machines *generating* electricity are known as dynamos.

1 Engines *which drive* dynamos usually derive their power from water or steam.
2 Electricity *which derives* its power from water is known as hydro-electricity.
3 Electric current *which flows* along a wire may be compared to water *which flows* along a pipe.
4 If a stone is dropped into a smooth lake, the surface is covered with circular waves *which move* outwards from the centre.
5 Parallel light rays *which strike* the surface of a concave mirror will converge when reflected.
6 Parallel light rays *which strike* the surface of a convex mirror will diverge when reflected.
7 Parallel light rays *which pass* through a concave lens will diverge.
8 Parallel light rays *which pass* through a convex lens will converge.
9 When parallel light rays have passed through a convex lens, the image *which appears* on the white card will be inverted.
10 Most substances consist of two or more elements *which combine* chemically as molecules to form compounds.
11 A body *which falls* freely in air accelerates at the rate of approximately 32 feet per second per second.
12 The bacteria *which combine* free nitrogen with other chemical elements are of importance to plants in building their tissues.

Exercise 4 Answer these questions without referring to the Texts:

1 Where is electricity generated?
2 What is the name of the large machines generating it?
3 What are they in fact?
4 From where do generators usually derive their driving power?
5 What is meant by the term 'hydro-electricity'?
6 What is the flow of electricity along a wire known as?
7 What is meant by 'electric potential'?
8 Which offers more resistance to the flow of water, a thin or a thick pipe?
9 Which offers more resistance to the current of electricity, a thin or a thick wire (of the same metal)?
10 Would you expect metals to be good or poor conductors of electricity?

Exercise 5 Questions for further discussion:

1 Give some examples of the uses of electricity.
2 Why is electricity cheaper in some countries than in others?
3 Have you seen small dynamos on bicycles? From where do they derive their power?
4 What ways of storing electricity do you know about? Where are these commonly used?
5 How could the current flowing along a wire be increased or decreased? (Think of water flowing along a pipe.)
6 Where can we see electricity in nature? Give examples.

Exercise 6 Suggestions for further activities:

1 Take a small battery, a small bulb, and two pieces of wire. Connect one wire to the positive and one to the negative pole of the battery. Put the other ends of the wires on to the base of the bulb, which will then light up.
2 Make induced electricity with a magnet using about two feet of insulated wire, a compass and a bar magnet. Wrap the insulated wire several times around your fingers and join the two ends, which should be uninsulated. Pass part of the wire over the compass. Push one pole of the bar magnet into the coil of wire and watch the action of the compass needle. Remove the bar magnet and observe the action of the compass needle. Repeat this using the opposite pole of the bar magnet. Is there any difference in the action of the compass needle this time?

Revision Exercises VII (Units 29–33)

I Give the meaning in your own language of these words:

1 neutral
2 protein
3 planet
4 to behave
5 charge
6 iron filings
7 negative
8 to mention
9 tilted
10 to sprinkle
11 path
12 the flow
13 steel
14 dynamo
15 chemical

II Explain the meaning of:

1 elliptical orbits
2 is magnetised
3 and vice versa
4 achieved by friction
5 at right-angles to
6 chemical formula
7 maintain their orbits
8 hydro-electricity
9 decomposed into other substances
10 electrically neutral
11 indicate the location
12 unlike poles
13 a magnetic field
14 as explained previously
15 electro-magnetism
16 may be compared to

III Give ONE word meaning:

1 produce (magnetism) without touching
2 taken away
3 knocked out of place
4 pulls towards itself
5 in this way, like this
6 the thing mentioned second
7 of gravity
8 only one
9 pushes away
10 often found
11 to look back to
12 to think of

IV Answer these questions without referring to the Texts:

1 What unit is used to measure electric current?
2 If two north poles of a magnet were placed near each other, what would they do?
3 Where does the main force of a bar magnet lie?
4 Does the earth travel round the sun, or the sun round the earth?
5 What causes the change (a) from day to night (b) of the seasons, on earth?
6 Why does the moon look bright at night?
7 What is meant by 'the solar system'?
8 What is a molecule?
9 What is the difference between an element and a compound?

10 If given the chemical formula of a molecule, can you say how many atoms it is composed of? (Give examples.)
11 If given the chemical formula of a substance, can you say whether it is an element or a compound? (Give examples.)
12 Name an element which has a large number of electrons in its atom.
13 How do electrons maintain their orbits?
14 If you know that an atom possesses three protons, how many electrons would you expect it to possess?
15 Does a whole atom carry a negative or a positive charge? (Give a reason for your answer.)
16 What is a simple method of achieving a static electric charge?
17 Which offers a greater resistance to electric current, a thin or a thick wire of the same metal?
18 What do we measure in amperes?
19 Of what does the nucleus of an atom consist mainly?

V Find the correct word or phrase with which to complete each of these sentences:

1 An —— is a substance which cannot be decomposed into other substances.
(a) organism (b) electron (c) element (d) atom
2 The simplest atom is that of hydrogen, which possesses a —— electron.
(a) complex (b) single (c) electric (d) elliptic
3 The planet most distant from the sun is —— .
(a) the moon (b) Mercury (c) Pluto (d) the earth
4 The most common compound is —— .
(a) water (b) air (c) uranium (d) carbon dioxide
5 Magnetism which is —— by an electric current is known as electro-magnetism.
(a) magnetised (b) induced (c) attracted (d) repelled
6 Like poles —— each other.
(a) attract (b) magnetise (c) modify (d) repel
7 The earth's —— is not at right-angles to its orbit, but is slightly tilted.
(a) planet (b) solar system (c) axis (d) balance
8 To explain how static electricity works, we must —— to the electrical balance of the atom.
(a) refer (b) maintain (c) equal (d) add
9 Some —— , such as proteins, possess hundreds of atoms of several elements in their molecules.
(a) elements (b) compounds (c) atoms (d) molecules

Unit 34 Life History of the Frog

Vocabulary

frog	insect	jelly	tadpole
gill			

A

The eggs of the frog consist of *small balls* of substance *like jelly,/with* a black spot in the centre. This black spot eventually becomes the tadpole.

First the black spot develops into a curved object, the tail *grows longer*, and the tadpole leaves the egg. Then it *hangs on* to a *plant living in water*
5 and remains there for some time. Later the mouth *can be seen*, and it begins to *eat* plants living in water. It breathes like a fish, *by using* gills which are *on the outside* at this *time*.

After approximately 6 weeks, the *back* legs appear *where the tail joins the body*, and soon the toes can be seen. The *front* legs do not appear
10 until approximately the 12th week.

By the end of the 8th week, however, the tadpole begins to take in air from the surface of the water, and the outside gills begin to disappear. *When it is three months old*, the mouth-*hole is larger*, and the tail has *grown shorter*. The tongue had grown longer, and it breathes by using
15 lungs. It is now ready to leave the water as a *fully developed* frog.

In the later part of the tadpole's life, it eats a small amount of animal food, such as *insects which live in water*.

Thus *a great change happens to the frog* during its development; i.e. it is an *animal which begins life in water but finishes it on land*.

B

(1)Frog-spawn consists of (2)globules of (3)jelly-like substance, (4)having a black spot in the centre, which eventually becomes the tadpole.

First the black spot develops into a curved object, the tail (5)lengthens, and the tadpole leaves the egg. Then it (6)adheres to a (7)water-plant and remains there for some time. Later the mouth (8)is evident, and it begins to (9)feed on (7)water-plants. It breathes like a fish, (10)by means of gills which are (11)external at this (12)stage.

After approximately six weeks, the (13)hindlegs appear (14)at the base of the tail, and soon the toes (8)are evident. The (15)forelegs do not appear until approximately the twelfth week.

By the end of the eighth week, however, the tadpole begins to take in air from the surface of the water, and the (11)external gills begin to disappear. (16)At three months, the mouth-(17)opening is enlarged, and the tail has (18)shortened. The tongue has (5)lengthened, and it breathes (10)by means of lungs. It is now ready to leave the water as an (19)adult frog.

(20)During the later stages of the tadpole's life, it (9)feeds on a small amount of animal food, such as (21)water insects.

Thus (22)the frog undergoes a metamorphosis during its development; i.e. it is an (23)amphibian.

Exercise 1 Find the way in which the words and phrases italicised in Text A are expressed in Text B:

1 The eggs of the frog
2 small balls
3 like jelly
4 with
5 grows longer
6 hangs on
7 plant living in water
8 can be seen
9 eat
10 by using
11 on the outside
12 time
13 back
14 where the tail joins the body
15 front
16 When it is three months old
17 hole is larger
18 grown shorter
19 fully developed
20 In the later part
21 insects which live in water
22 a great change happens to the frog
23 animal which begins life in water but finishes it on land

Exercise 2 Some verbs are formed by adding EN-/-EN:

(1a) to *become longer* (1b) to *lengthen*
(2a) to *get larger* (2b) to *enlarge*

Using a verb formed with EN-/-EN in place of each italicised phrase, as in examples (b), read or rewrite these sentences:

1 The black spot *increases in size* and *gets loose* in the jelly.
2 The tail of the tadpole *grows longer*, and the tadpole leaves the egg.
3 At three months, the mouth-opening *becomes larger*, and the tail *grows shorter*.
4 The body of the tadpole also *grows thicker*.
5 When we plant seeds, we must *make sure* that they are not too crowded.
6 If bean seeds are soaked in water, they *become softer*.
7 When the metal band around the wheel contracts on cooling, it *gets tighter*.
8 In order to increase the current, the wire should be *made thicker*, and this will *make* the resistance *less*.
9 By *making* the wire *thicker*, the resistance will be *made less*.
10 In order to *make* the current *stronger*, the potential must be increased.

Exercise 3 The ten drawings of the Development of the Frog are in the correct order, but the ten sentences given below are not.

(i) Write the appropriate sentence under each drawing.
(ii) Label the drawings with the words or phrases italicised in the sentences.

(a) The *hind-legs* appear at the *base of the tail* at six weeks.
(b) After leaving the egg, it adheres to a *water-plant*.
(c) The *mouth-opening* enlarges, the *tail* shortens, the *tongue* has lengthened and there are no external gills.
(d) The *mouth* becomes evident and it breathes by means of *external gills*.
(e) It is ready to leave the water as an *adult frog*.
(f) The black spot develops into a *curved object*.
(g) The *fore-legs* appear at approximately twelve weeks.
(h) *Frog-spawn* consists of *globules* of jelly-like substance having a *black spot* at the centre.
(i) The *external gills* begin to disappear and it begins taking in air from the *surface*, at eight weeks.
(j) The *tail* lengthens and the *tadpole* leaves the *egg*.

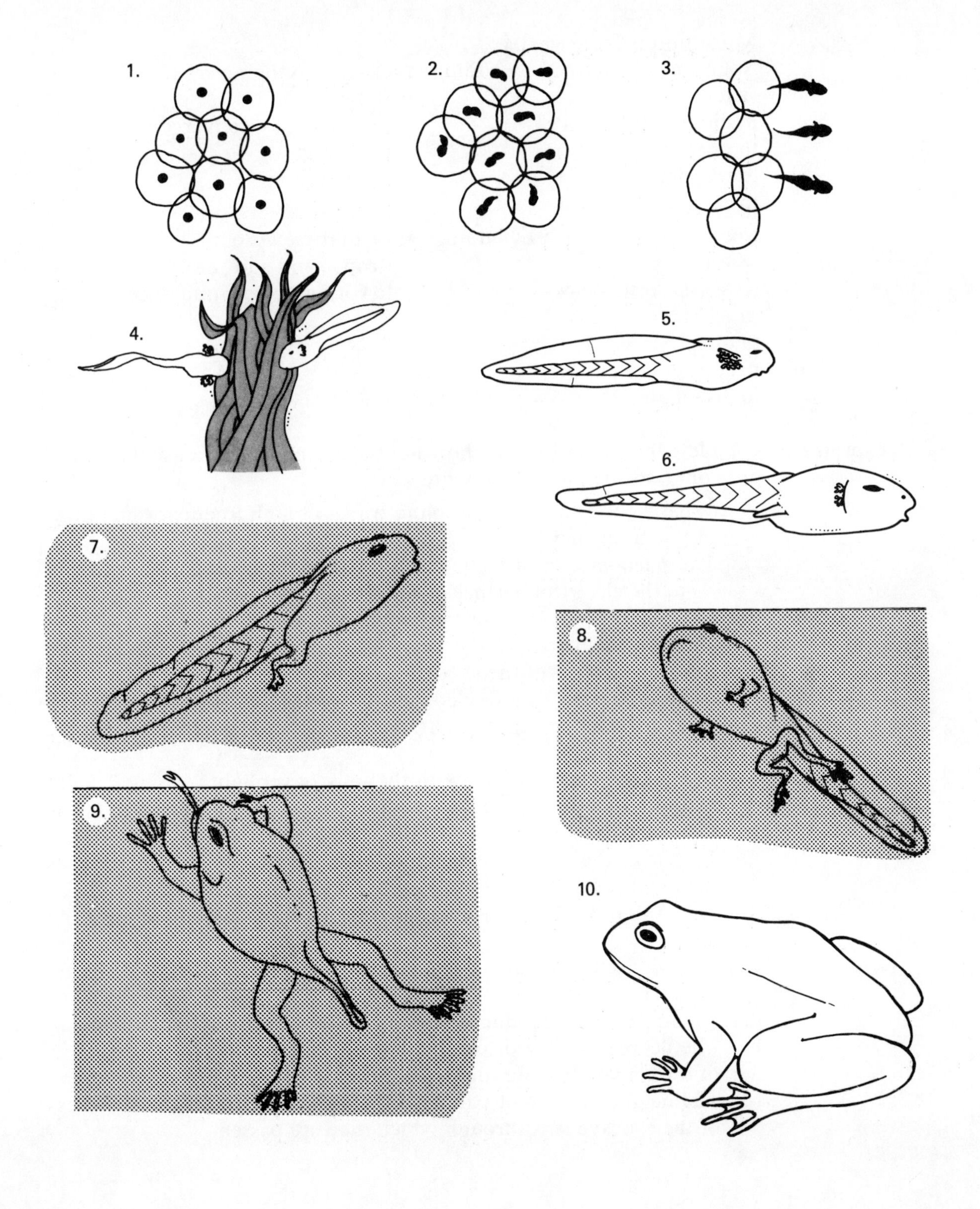
1.
2.
3.
4.
5.
6.
7.
8.
9.
10.

More nouns from verbs

Nouns may be formed by adding -ENCE/-NCE to these verbs:

-ENCE	-NCE
to differ – differ-ence	to converge – converge-nce
to refer	to diverge
to transfer	to emerge
to exist	to adhere

Nouns may be formed by adding -ANCE to these verbs:

-ANCE	Drop -e and add -ANCE
to resist – resist-ance	to continue – continu-ance
to assist	
to disturb	
to appear	
to disappear	

Exercise 4 Complete the above table as shown in the examples. Then use the nouns, once only, to complete these sentences:

1 What is the —— between sound waves of high frequency and those of low frequency?
2 The phenomenon of static electricity cannot be explained without — to the electrical balance of the atom.
3 The —— of heat in metals is known as conduction.
4 First the main root of the bean seed appears, and then we notice the gradual —— of the shoot.
5 At approximately six weeks, the —— of the hind-legs can be noted.
6 At approximately eight weeks, we may note the gradual —— of the external gills.
7 Green leaves make starch with the —— of sunlight.
8 A thin wire offers more —— to the current than a thick one of the same metal.
9 To ensure the —— of their kind, plants produce many more seeds than will survive.
10 The —— of seeds to the coats of animals is one method of ensuring dispersal.
11 The —— of life on earth depends completely on the sun, directly and indirectly.
12 Sound waves are produced by the —— of the molecules of a medium.
13 The degree of —— of parallel light rays depends on the thickness of the convex lens through which they are passed.
14 The degree of —— of parallel light rays depends on the thickness of the concave lens through which they are passed.

Exercise 5 Answer these questions without referring to the Texts:

1 What does frog-spawn consist of?
2 What does the black spot first develop into?
3 What does the tadpole do when it first leaves the egg?
4 What does it feed on first?
5 How does it breathe at this stage?
6 What appear at approximately 6 weeks?
7 When does the tadpole first begin to take in air?
8 What begin to disappear at this stage?
9 At three months, what has (a) lengthened? (b) shortened? (c) enlarged?
10 What does it feed on at this stage?
11 How does it breathe at this stage?
12 What is meant by the term 'amphibian'?

Exercise 6 Questions for further discussion:

1 Is a frog warm- or cold-blooded? What other animals do you know to be cold-blooded?
2 The amphibian is a stage of development between the fish and the reptile. Give reasons why you think this is true.
3 Why do you think one pair of frogs produces so many eggs in one season?
4 Do you know of any amphibian besides the frog?
5 What do frogs do (a) in winter? (b) in summer?
6 What are some of the differences and similarities between a frog, a fish, and a lizard?

Vocabulary

lizard reptile

Unit 35 Digestion in Humans

Vocabulary

anus	intestine	muscle	stomach
to digest	juices	nourishment	to swallow
digestion	liver	pancreas	

A

Food is digested as it passes along the *long tube which begins at the mouth and ends at the anus*. The process of digestion begins in the mouth where the food is *crushed by the teeth* and *made wet* by the *juices in the mouth*. After *it is* swallowed, it passes down the gullet, a tube *having muscles*
5 and *going* to the stomach.

The stomach walls, which have muscles, break up the food *still more*. The *juices of the stomach*, which are *sent out* from the *inner layer of the stomach wall*, continue the process of digestion which was started in the mouth.

10 The food is then passed into the small intestine. The upper *part/gets* the juices sent out by the liver and pancreas. These juices convert the food still more, so that it can be *taken in and used* by the body.

The food passes from the middle part of the small intestine into the lower part, *whose walls* are covered with tiny *things that stick out/and look*
15 *like* hairs. These are known as villi. Their function is to *take in* the nourishment from the food.

In the large intestine, *no more food is digested after this*. Most of the liquid is taken in from the material *which has not been digested*, and the remainder is *thrown out of the body* as *material which cannot be used*.

B

Food is digested as it passes along the (1)alimentary canal. The process of digestion begins in the mouth where the food is (2)chewed and (3)moistened by the (4)saliva. After (5)being swallowed, it passes down the gullet, a (6)muscular tube (7) leading to the stomach.

The (6)muscular stomach walls break up the food (8)further. The (9)gastric juices, which are (10)secreted from the (11)gastric lining, continue the process of digestion which was started in the mouth.

The food is then passed into the small intestine. The upper (12)section (13)receives the juices (10)secreted by the liver and pancreas. These juices (8)further convert the food, so that it can be (14)assimilated by the body.

The food passes from the middle (12)section of the small intestine into the lower (12)section, (15)the walls of which are covered with tiny (16) projections (17)resembling hairs. These are known as villi. Their function is to (18)absorb the nourishment from the food.

In the large intestine, (19)no further digestion takes place. Most of the liquid is (18)absorbed from the (20)undigested material, and the remainder is (21)excreted as (22)waste.

Exercise 1 Find the way in which the words and phrases italicised in Text A are expressed in Text B:

1 long tube which begins at the mouth and ends at the anus
2 crushed by the teeth
3 made wet
4 juices in the mouth
5 it is
6 having muscles
7 going
8 still more
9 juices of the stomach
10 sent out
11 inner layer of the stomach wall
12 part
13 gets
14 taken in and used
15 whose walls
16 things that stick out
17 and look like
18 take in
19 no more food is digested after this
20 which has not been digested
21 thrown out of the body
22 material which cannot be used

Exercise 2 Compound nouns: Rewrite the following phrases as compound nouns, as shown in example (b):

Phrase	Compound noun:
(a) *the wall of the stomach*	(b) *stomach wall*

1 the juices of the stomach
2 the lining of the stomach
3 the material which is waste
4 the spawn of the frog
5 a plant which lives in water
6 insects which live in water
7 animals which live on land
8 the opening of the mouth
9 the length of the wave
10 a molecule of water
11 an image in a mirror
12 the motion of waves
13 waves of light
14 waves of sound
15 waves of heat
16 the pressure of water
17 the pressure of air
18 impulses of the nerves
19 the transfer of heat
20 the medium of sounds
21 resistance of air
22 conductor of heat
23 needle of a compass
24 a molecule of protein

Exercise 3 Rewrite this passage, using passive forms. (The subjects of the passive sentences are italicised). You will then have made a summary of the Texts:

The teeth crush *the food* and the saliva moistens *it*. The muscular walls of the stomach break *it* down still further, while the gastric lining secretes *juices* to continue the process. The liver and pancreas secrete *juices* into the upper section of the smaller intestine, and these convert *the food* so that the body can assimilate *it*. In the intestine, tiny projections cover *the walls*, and these absorb *the nourishment* from the food. Here the body absorbs *most of the liquid* from the undigested food, and excretes *the remainder* as waste.

Diagram of alimentary canal

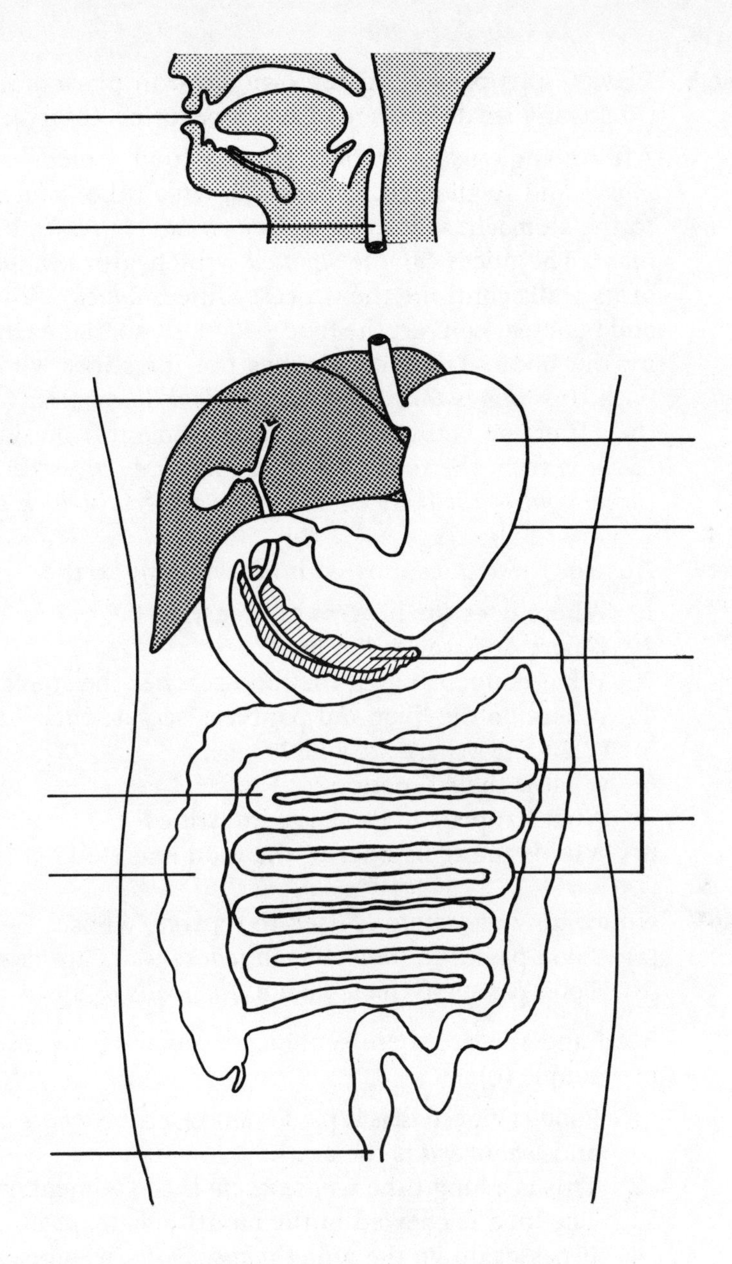

Exercise 4 On the simple diagram of the alimentary canal, write the names of the parts shown by the arrows:

1 mouth
2 gullet
3 stomach
4 stomach lining
5 liver
6 pancreas
7 small intestine
8 villi
9 large intestine
10 anus

Exercise 5 Rewrite this passage, using ONE word in place of each phrase italicised. (Make any small *change of word order necessary):

After being *crushed by the teeth*, the food is *made wet* by the *juices in the mouth* and swallowed. It passes down a tube **which has muscles/and goes* to the stomach, whose walls **which have muscles* break up the food *still more*. The juices **of the stomach* which are *sent out* by the *inner layer* of its walls continue the process. Juices *which are sent out* from the liver and pancreas convert the food *still more*, so that it can be *taken up and used* by the body. The food reaches the intestine, whose walls are covered with tiny *things that stick out and look like* hairs. These are the villi and their function is to *take in* the nourishment from the food. Liquid is then *taken in* from the food **which has not been digested*, and the remainder is *thrown out of the body* because it is *material which cannot be used*.

Exercise 6 Answer these questions without referring to the Texts:

1 Where does the process of digestion begin?
2 Where does the gullet lead?
3 What happens when the food reaches the stomach?
4 Where do the liver and pancreas secrete their juices?
5 What and where are villi?
6 What is their function?
7 What happens in the large intestine?
8 Why is the remainder of the food excreted?

Exercise 7 Notice how OF WHICH is used to replace 'whose':

(a) Food passes to the stomach *whose walls* are muscular.
(b) Food passes to the stomach, *the walls of which* are muscular.

Read and rewrite these sentences, replacing 'whose' with OF WHICH, as in example (b):

1 Food is digested as it passes along a tube *whose beginning* is the mouth and *whose end* is the anus.
2 This is a long tube *whose name* is the alimentary canal.
3 The food is chewed in the mouth *whose juices* moisten it.
4 It passes down the gullet *whose walls* are muscular.
5 It is further broken down in the stomach *whose lining* secretes juices.
6 The stomach continues the process of digestion *whose first stages* took place in the mouth.
7 The food passes to the small intestine *whose upper section* receives juices secreted by the liver and pancreas.
8 The liver and pancreas, *whose juices* are received into the small intestine, further aid the process of digestion.

9 Then the food passes into the lower section *whose walls* are covered with tiny projections.
10 The villi, *whose function* is to absorb nourishment, cover the walls of the lower section.
11 The food passes into the large intestine *whose function* is to absorb most of the liquid from the food.

Exercise 8 Questions for further discussion:

1 Why is food necessary?
2 Why is digestion necessary?
3 Why do babies have to drink only milk in the first few months of their life?
4 Why do people with liver trouble need a special diet?
5 What are some of the differences between the diets of (a) a lion, (b) a sheep and (c) a man?
6 In what way do the teeth of these animals show us that they have different eating habits?
7 Why does man need three meals a day, whereas a sheep needs to eat nearly all the time, and a lion need to eat only two or three times a week?
8 Why isn't wood a food for man, as it is for some insects?
9 What are some of the main differences between human eating habits and those of all other animal life?

Vocabulary
diet

Unit 36 Solutions, Suspensions and Colloids

Vocabulary

clear microscope to stir
flour

A

If you add some sugar to water and stir it, the sugar disappears; in other words, it dissolves, because sugar is *a substance which dissolves* in water. It is said to have formed a solution, with sugar as the *substance which has been dissolved*.

5 If you add some flour to water and stir it, the flour does not disappear; in other words, it does not dissolve, because flour is *a substance which does not dissolve* in water. Instead of dissolving, the *little pieces* of flour *hang about* in the water *and make it look 'milky'*. We say that they are suspended in the water; in other words, they have formed a suspension. 10 A suspension is a *scattering* of little pieces of a solid, *which can be seen under a microscope*, in a *liquid or gas*.

Gradually, the little pieces of flour begin to *sink to the bottom*, until they are deposited at the bottom of the water, which becomes clear again. *Small pieces sink more slowly; large pieces sink more quickly*.

15 The little pieces in a suspension *can be seen/by using* a microscope. But if the *little pieces hanging about* are so small that they would never sink, and *cannot be seen* even by using a microscope (e.g. protein molecules in the blood), they are said to have formed a colloid. A colloid is a suspension of little pieces of a solid *which cannot be seen under an ordinary microscope*.

20 *The difference between a suspension and a colloid is* in the size of the little pieces hanging about in the liquid or gas. The *little pieces of a colloid* are charged with electricity, most of them negatively.

B

If some sugar is added to water and stirred, the sugar disappears; in other words, it dissolves, because sugar is (1)soluble in water. It is said to have formed a solution, with sugar as the (2)solute.

If some flour is added to water and stirred, the flour does not disappear;
5 in other words, it does not dissolve, because flour is (3)insoluble in water. Instead of dissolving, the (4)particles of flour (5)are suspended in the water, (6)giving it a 'milky' appearance. They are said to be suspended in the water; in other words, a suspension has been formed. A suspension is a (7)dispersion of (8)microscopic (4)particles in a (9)fluid.

10 Gradually, the (4)particles of flour begin to (10)settle, until they are deposited at the bottom of the water, which becomes clear again. (11)The smaller the (4)particles, the slower they (10)settle; the larger the (4) particles, the quicker they (10)settle.

The (4)particles in a suspension (12)are visible (13)with the aid of a
15 microscope. But if the (14)suspended particles are so small that they would never (10)settle, and (15)are invisible even (13)with the aid of a microscope (e.g. protein molecules in the blood), they are said to have formed a colloid. A colloid is a suspension of (16)submicroscopic (4)particles.

(17)A suspension differs from a colloid in the size of the (14)particles
20 suspended in the (9)fluid. (18)Colloidal particles are charged with electricity, most of them negatively.

Exercise 1 Find the way in which the words and phrases italicised in Text A are expressed in Text B:

1 a substance which dissolves
2 substance which has been dissolved
3 a substance which does not dissolve
4 little pieces
5 hang about
6 and make it look 'milky'
7 scattering
8 which can be seen under a microscope
9 liquid or gas
10 sink to the bottom
11 Small pieces sink more slowly; large pieces sink more quickly
12 can be seen
13 by using
14 little pieces hanging about
15 cannot be seen
16 which cannot be seen under an ordinary microscope
17 The difference between a suspensions and a colloid is
18 little pieces of a colloid

Exercise 2 HAVING is often used to replace 'with/which has':

(a) A colloid is a suspension *which has* submicroscopic particles in it.

(b) A colloid is a suspension *having* submicroscopic particles in it.

Read of rewrite these sentences as in example (b):

1 Frog-spawn consists of globules of jelly-like substance with a black spot in the centre.
2 The tadpole, which has external gills, at first breathes like a fish.
3 The stomach may be compared to a bag which has a muscular wall.
4 The whole atom is electrically neutral, with an equal number of protons and electrons.
5 The simplest atom is that of hydrogen, which has a single electron.
6 An atom of uranium possesses a larger nucleus with 92 protons.
7 The chemical formula for water, which has 2 atoms of hydrogen and 1 of oxygen in each of its molecules, is H_2O.
8 The chemical formula for carbon dioxide, which has 2 atoms of oxygen and 1 of carbon in each of its molecules, is CO_2.
9 The chemical formula for potassium permanganate, which has 1 atom of potassium (K), 1 of manganese (Mn) and 4 of oxygen (O) in each of its molecules, is $KMnO_4$.
10 A complex atom possesses a large nucleus with a greater number of protons in it.

Exercise 3 IS SAID TO HAVE + past participle: Look at these two sentences:

(1a) *We say that* it *has dissolved.* (2a) *We say that* the sugar *has dissolved.*

In scientific writing, the (b) forms given below are preferred:

(1b) It *is said to have dissolved.* (2b) The sugar *is said to have dissolved.*

Read or rewrite these sentences as in examples (b):

1 We say that the sugar has formed a solution.
2 We say that the flour has formed a suspension.
3 We say that the protein molecules have formed a colloid.
4 When the suspended particles sink to the bottom, we say that they have settled.
5 If a piece of iron begins to attract small pieces of iron, we say that it has been magnetised.
6 When everything else has been extracted, we say that a vacuum has been formed.
7 When a tadpole has changed into a frog, we say that it has undergone a metamorphosis.
8 When free nitrogen has been chemically combined, we say that it has been fixed.
9 When iron filings stick to a magnet, we say that they have been attracted.
10 When two or more elements have formed a molecule, we say that they have chemically combined.
11 When two or more elements have chemically combined, we say that they have formed a compound.
12 When food has passed along the alimentary canal, we say that it has been digested.

Exercise 4 Rewrite this passage, using passive forms (with agent as necessary). (The subjects of the passive sentences are italicised):

If you add *some sugar* to water and stir *it*, the sugar dissolves, and we call *it* the solute. We can say that *it* has formed a solution.

If you add *some flour* to water and stir *it*, the flour does not dissolve. We say that *it* has formed a suspension. We can see *the tiny particles of a suspension* with the aid of a microscope, but we cannot see *the tiny particles of a colloid*, even with a microscope. Most colloidal particles carry *a negative electric charge*.

Exercise 5 BY THIS IS MEANT is used to show that an idea will be repeated in different words:

(a) The whole atom is electrically neutral. The whole atom has no electric charge.

These sentences can be joined together like this:

(b) The whole atom is electrically neutral; *by this is meant that* the whole atom has no electric charge.

Find the sentence in List B which 'matches' a sentence in List A, and join them together as in example (b):

List A

1 Some water is said to be hard.
2 When hard water is boiled, layers of fur are left behind.
3 The weight of a body may equal the weight of water it displaces.
4 When warm humid air meets a cold surface, the water vapour condenses.
5 When water is sufficiently cooled, it solidifies.
6 A given volume of water increases on becoming ice.
7 A seed must find the right conditions if it is to live.
8 A candle goes out when it has used up all the oxygen in the bell jar.
9 In the blood, sugar is consumed.
10 Vegetation formed carbohydrates with the aid of sunlight.
11 Space remains cold although the sun's rays are passing through it.
12 Like poles repel and unlike poles attract.
13 The heat is transferred readily from one molecule to the next.
14 Gases quickly increase in volume when they are heated.

List B

(a) Water expands on solidifying.
(b) The sun's energy was converted into chemical energy.
(c) It turns to ice.
(d) Calcium compound is deposited.
(e) Electro-magnetic waves travel through a vacuum.
(f) The remaining gases will not support combustion.
(g) A state of equilibrium will then be reached.
(h) All metals are good heat conductors.
(i) It returns to the liquid state.
(j) It cannot survive in an unsuitable environment.
(k) On heating, they expand rapidly.
(l) It does not lather readily with soap.
(m) Oxygen is replaced by carbon dioxide, just as when a candle burns.
(n) Two north poles repel, and a north pole and a south pole attract each other.

Exercise 6 Read or rewrite this summary, using a single word in place of each phrase italicised. (One *change in word order is necessary):

Sugar is *a substance which dissolves* in water, but flour is *a substance which does not dissolve* in water. The *tiny pieces* of flour form a suspension in the water until they gradually *sink to the bottom*. A suspension is a dispersion of *little pieces*, **which can be seen under a microscope*, in a *liquid or gas*. The *tiny pieces* in a colloid, however, are *invisible under the microscope*, and they are so small that they will never *sink to the bottom*. Most colloidal particles are charged *with negative electricity*.

Exercise 7 Answer these questions without referring to the Texts:

1 How is a sugar solution made?
2 What do we call the substance which has been dissolved to form a solution?
3 Is flour soluble or insoluble in water?
4 What do the flour particles do, instead of dissolving in the water?
5 What does a suspension of flour look like?
6 Which settle quicker, the large or the small particles in a suspension?
7 When do colloidal particles settle?
8 Which particles can be seen with a microscope, suspended or colloidal?
9 What is the main difference between a suspension and a colloid?
10 With what are the particles of most colloids charged?

Exercise 8 Questions for further discussion:

1 What do you think each of these substances might be, a solution, a suspension, or a colloid?

(a) muddy water	(d) blood	(g) glue	(j) ink
(b) water-paint	(e) lemonade	(h) sea-water	(k) smoke
(c) dust in air	(f) coca cola	(i) fog	

2 Do you know of any type of microscope with which you might be able to see the particles of some colloids?

Vocabulary
glue

Exercise 9 Suggestions for further activities:

Fill a small bottle with water and add a little flour. Shake it well and place the bottle on a table and wait for the particles to settle. See how long it takes for the water to become clear again. Repeat the experiment using different insolubles, such as sand, coal-dust, cigarette ash. Which takes longest to settle? Why?

Unit 37 Acids, Bases and Salts

Vocabulary

dye
equation
opposite
purple
to share
sour

A

Substances possessing a sour taste all contain compounds known as acids. All acids *share the same/characteristics*: (a) they all taste sour, (b) they turn *litmus paper red, (c) *something happens when they meet* most metals and they *set free* hydrogen gas, and (d) when *they are dissolved in water*, they
5 set free positively-charged particles known as hydrogen ions. An acid *which is often found* is hydrochloric acid (*written like this in chemistry*: HCl).

The substances possessing opposite characteristics to *the characteristics* of acids, are known as bases. If a base is soluble in water, it is called an alkali. Bases turn *litmus paper blue, and when they are dissolved in water,
10 they set free hydroxyl ions. A hydroxyl ion is *made up* of an atom of oxygen *joined together* with an atom of hydrogen. *Because* it carries one extra electron, it is negatively charged. A base which is often found is sodium hydroxide (written like this in chemistry: NaOH).

Because opposite electric charges attract each other, positive hydrogen
15 ions join together with negative hydroxyl ions and *make neutral* their charges, *so that they form* water. The equation *of ions* for this *happening* is: $H^+ + OH^- = H_2O$.

A substance known as a salt is also formed during this happening;
20 the equation *is* written like this in chemistry: $HCl + NaOH \rightarrow NaCl + H_2O$. This equation *is a way of saying* that an acid joins together with a base, so that a salt and water are formed. All salts are made up of a metal joined together with a *substance which is not a metal*, or with an acid *group of atoms which behaves like a single atom*.

B

Substances possessing a sour taste all contain compounds known as acids. All acids (1)possess (2)properties (1)in common: (a) they all taste sour, (b) they turn *litmus paper red, (c)(3)they react with most metals and (4)release hydrogen gas, and (d) when (5)in solution, they (4)release
5 positively-charged particles known as hydrogen ions. A (6)common acid is hydrochloric acid ((7)chemical formula: HCl).

The substances possessing opposite (1)properties to (8)those of acids, are known as bases. If a base is soluble in water, it is called an alkali. Bases turn *litmus paper blue, and when (5)in solution, they (4)release hydroxyl
10 ions. A hydroxyl ion is (9)composed of an atom of oxygen (10)combined with an atom of hydrogen. (11)Since it carries one extra electron, it is negatively charged. A (6)common base is sodium hydroxide ((7)chemical formula: NaOH).

(11)Since opposite electric charges attract each other, positive hydrogen
15 ions (10)combine with negative hydroxyl ions and (12)neutralize their charges (13)to form water. The (14)ionic equation for this (15)reaction is: $H^+ + OH^- = H_2O$.

A substance known as a salt is also formed during this (15)reaction, the chemical equation (16)being: $HCl + NaOH \rightarrow NaCl + H_2O$. This equation
20 (17)expresses the fact that an acid (10)combines with a base (13)to form a salt and water. All salts are (9)composed of a metal (10)combined with a (18)non-metal, or with an acid (19)radical.

*Paper which is coloured purple with a special dye called litmus.

Exercise 1 Find the way in which the words and phrases italicised in Text A are expressed in Text B:

1 share the same
2 characteristics
3 something happens when they meet
4 set free
5 they are dissolved in water
6 which is often found
7 written like this in chemistry
8 the characteristics
9 made up
10 joined together
11 Because
12 make neutral
13 so that they form
14 of ions
15 happening
16 is
17 is a way of saying
18 substance which is not a metal
19 group of atoms which behaves like a single atom

Exercise 2 THOSE can often be used to avoid repeating a plural noun:

(a) The simplest atoms are *the atoms* of hydrogen.
(b) The simplest atoms are *those* of hydrogen.

Read and rewrite these sentences, using THOSE to avoid repeating the nouns italicised:

1 The *compounds* of some proteins possess hundreds more atoms than the compounds of water or salt.
2 Substances possessing the *properties* opposite to the properties of acids are known as bases.
3 The *ions* of hydrogen are positively charged, but the ions of hydroxyl are negatively charged.
4 The *ions* of hydrogen combine with the ions of hydroxyl and neutralise their charges.
5 The *charges* of hydrogen ions are positive, but the charges of hydroxyl ions are negative.
6 The *orbits* of electrons may be compared to the orbits of the planets.
7 The *particles* in a suspension are larger than the particles of a colloid.
8 In other words, the *suspended particles* in a colloid differ in size from the suspended particles of a suspension.
9 The *molecules* of some compounds are larger than the molecules of others, and the *atoms* of some elements are larger than the atoms of others.
10 If you are moving forwards in a train, *objects* inside the train appear to be still, while objects outside appear to move backwards.
11 The *orbits* of the planets and the orbits of electrons are not circular but elliptical.
12 The *juices* excreted by the liver and the juices excreted by the pancreas further convert the food in the small intestine.
13 *Certain bacteria* found on the roots of plants and certain bacteria living in the soil combine free atmospheric nitrogen.
14 *Waves* of light travel at a much greater velocity than waves of sound.
15 *Parallel light rays* passing through a convex lens converge, and parallel light rays passing through a concave lens diverge.

Exercise 3 A simple infinitive is often used to replace the idea of 'so that/with the result that':

(a) Positive and negative ions combine *with the result that water is formed.*
(b) Positive and negative ions combine *to form water.*

Rewrite these sentences as in example (b):

1 An acid combines with a base with the result that a salt and water are formed.
2 Acids react with most metals with the result that hydrogen gas is released.
3 Sugar dissolves in water with the result that a solution is made.
4 Flour particles can be added to water and stirred so that a milky appearance is given.
5 Flour particles are suspended in the water with the result that not a solution, but a suspension, is formed.
6 An element is a substance which cannot be decomposed so that other substances are formed.
7 Two atoms of hydrogen and one of oxygen combine so that water is formed.
8 One atom of sodium, one of oxygen and one of hydrogen combine with the result that sodium hydroxide is formed.
9 Gastric juices are secreted from the stomach lining so that the process of digestion is continued.
10 A metal and a non-metal combine with the result that a salt is formed.
11 A group of atoms behaves like a single atom with the result that a radical is formed.
12 Two or more elements combine chemically as molecules so that compounds are formed.

Exercise 4 NON- may be added to some words, giving them the negative meaning of 'which is not/which does not':

(1a) a *substance which is not a metal* (1b) a *non-metal*
(2a) *which does not condense* (2b) *non-condensing*

Make the word meaning:

1 a substance which is not a conductor
2 which does not reflect
3 which is not poisonous
4 which is not living
5 which is not spherical
6 which does not adhere
7 which is not transferable
8 which is not combustible
9 which is not a radical
10 which is not electric
11 which is not chemical
12 which is not resistant
13 a substance which is not acid
14 a substance which is not a protein
15 which does not vibrate
16 which does not exist
17 which does not rotate
18 which is not absorbable

Exercise 5 Read or rewrite this summary, using more 'scientific' terms in place of the words and phrases italicised:

Acids *share the same characteristics*: (a) they taste sour, (b) they turn litmus paper red, (c) they react with most metals *so that they set free* hydrogen gas, and (d) when *they are dissolved in water*, they *set free* positively-charged hydrogen ions.

Bases are substances having opposite *characteristics*. They turn litmus paper blue, and when *dissolved in water*, they *set free* hydroxyl ions which are negatively charged, *because* they carry one extra electron.

Hydrogen ions *join together* with hydroxyl ions *so that they form* water and a salt. A salt is *made up* of a metal *joined together* with a *substance which is not a metal*, or with an acid *group of atoms which behaves as a single atom*.

Exercise 6 Answer these questions without referring to the Texts:

1 How many properties have acids in common?
2 What happens when an acid reacts with a metal?
3 What is released when acids are in solution?
4 What is a hydrogen ion?
5 What is the substance whose chemical formula is HCl?
6 What colour is litmus paper? What colour does it turn when dipped into (a) an acid? (b) an alkali?
7 What is (a) a base? (b) an alkali?
8 What is a hydroxyl ion?
9 Is NaOH an acid or a base?
10 What fact is expressed by the equation: $H^+ + OH^- = H_2O$?
11 When a metal and a non-metal combine, what is formed?
12 What is meant by 'a radical'?

Exercise 7 Questions for further discussion:

1 Are any acids manufactured in your country? If so, can you find out where and what?
2 Can you discover which bases are used in the manufacture of soap?

Vocabulary

baking powder   detergent   toothpaste   vinegar

Exercise 8 Suggestions for further activities:

1 Get some *litmus paper (from a laboratory or a chemist's shop) and, by dipping it into the following, find out whether they are acid, alkali, or neutral:

(a) vinegar
(b) your own saliva, both before and after a meal
(c) toothpaste
(d) a solution of starch
(e) salt-water
(f) soapy water
(g) water in which soil has been soaked
(h) drinking water
(i) coca cola
(j) liquid detergent
(k) lemon juice
(l) tea
(m) coffee
(n) milk
(o) yogurt (or sour milk)
(p) sugar solution
(q) tomato juice

* If red and blue litmus papers are used, the liquids must be tested with both, i.e. with an acid, the red stays red, but the blue turns red; with an alkali, the blue stays blue, but the red turns blue; with a neutral, the red stays red and the blue stays blue.

2 Drop a few drops of vinegar onto some baking powder, and describe what happens. Can you explain why? (Chemical formula of baking powder is $NaHCO_3$, and for vinegar it is HCl.)

Revision Exercises VIII (Units 34–37)

I Give the meaning of these words in your own language:

1 juices
2 muscle
3 frog
4 nourishment
5 purple
6 insect
7 gill
8 flour
9 grain
10 tadpole
11 to stir
12 intestine
13 equation
14 to swallow
15 to digest
16 liver
17 opposite
18 stomach
19 sour
20 dye

II Explain the meaning of:

1 feeds on water plants
2 soluble in water
3 becomes evident
4 undergoes a metamorphosis
5 when in solution
6 an amphibian
7 resembling hairs
8 possess four properties in common
9 alimentary canal
10 there is no further digestion
11 submicroscopic particles
12 react with most metals
13 external at this stage
14 a radical
15 giving it a milky appearance

III Give ONE word meaning:

1 a liquid or a gas
2 joined together
3 can be seen
4 things that stick out
5 crushed by the teeth
6 make neutral
7 substance which has been dissolved
8 set free
9 grown shorter
10 which has muscles
11 material which cannot be used
12 sink to the bottom
13 substance which is not a metal

IV Answer these questions without referring to the Texts:

1 Approximately how long does it take the frog to develop from its spawn?
2 If a tadpole were taken out of the water, would it die? (Give a reason for your answer.)
3 Where does digestion in humans (a) take place? (b) start? (c) end?
4 What are the villi? Where are they found and what is their function?

5 What fact is expressed in this equation: $HCl + NaOH \rightarrow H_2O + NaCl$?
6 What is a suspension?
7 What is the difference between (a) a solution and a suspension? (b) a suspension and a colloid?
8 What happens to the body's waste material?
9 Which organs secrete their juices into the small intestine?
10 What is the name of the muscular tube leading from the mouth to the stomach?
11 Which settle faster, the smaller or the larger particles in a suspension?
12 What colour is litmus paper? What colour is it when dipped into (a) an acid? (b) an alkali?
13 What is released when an acid reacts with a metal?
14 What is released when an acid is in solution?
15 What is released when a base is in solution?
16 What is meant by the term 'an alkali'?
17 Why do hydrogen ions and hydroxyl ions combine readily?
18 What two substances are formed when an acid and an alkali combine?
19 What does frog-spawn look like?
20 Which appear first, the hind- or the fore-legs of the tadpole?
21 What is the difference in the breathing of the tadpole and the frog?

V Find the correct word or phrase with which to complete each of these sentences:

1 Juices in the small intestine further —— the food so that it can be assimilated by the body.
(a) moisten (b) continue (c) secrete (d) convert
2 If a substance disappears when added to water, it is said to be —— in water.
(a) soluble (b) suspended (c) deposited (d) insoluble
3 Juices which are —— by the gastric lining continue the process of digestion.
(a) absorbed (b) excreted (c) moistened (d) secreted
4 If some flour is added to water and stirred, it can be said to have formed a ——.
(a) solution (b) suspension (c) colloid (d) deposit
5 All acids possess four —— in common.
(a) ions (b) reactions (c) properties (d) charges
6 When in solution, —— turn litmus paper blue.
(a) bases (b) salts (c) acids (d) ions

7 A —— ion consists of an atom of hydrogen combined with an atom of oxygen.
(a) hydrochloric acid (b) non-metal (c) hydroxyl (d) hydrogen

8 A group of atoms which behaves as a single atom is known as a ——.
(a) molecule (b) radical (c) hydroxide (d) non-metal

9 A hydroxyl ion carries a negative charge because it possesses one extra ——.
(a) proton (b) neutron (c) electron (d) nucleus

10 Positive hydrogen ions combine with negative hydroxyl ions and thus —— their charges.
(a) release (b) attract (c) express (d) neutralize

11 A suspension is a —— of microscopic particles in a fluid.
(a) dispersion (b) function (c) process (d) deposit

12 When a tadpole first leaves the egg, it —— to a water-plant for some time.
(a) remains (b) feeds on (c) undergoes (d) adheres

VI Complete each of these sentences with a suitable noun ending in -ION:

1 Things which project are known as ——.
2 A number which approximates (i.e. is approximately correct) is known as an ——.
3 The process of moving is known as ——.
4 The process of absorbing is known as ——.
5 The process of acting is known as ——.
6 The process of reacting is known as ——.
7 The process of secreting is known as ——.
8 The process of digesting is known as ——.

VII Rewrite these passages using passive forms, with or without agent as necessary:

(a) In the mouth, the saliva moistens the food. When the food passes into the stomach, the gastric juices continue the process of digestion. The pancreas and the liver secrete juices into the upper section of the small intestine. The juices further convert the food so that the body can assimilate it. Tiny projections cover the walls of the lower section of the small intestine. We call these the villi. They absorb the nourishment from the food. The large intestine absorbs most of the liquid from undigested material and excretes the remainder as waste.

(b) If you add sugar to water, it forms a solution and we call the sugar the solute. If you add flour to water it does not dissolve, but we say we have formed a suspension. We can see the particles of a suspension with the aid of a microscope. If, however, particles suspended in a fluid are invisible even under a microscope, we say that they have formed a colloid. Most colloidal particles carry a negative charge.

(c) When an acid reacts with a metal, it releases hydrogen gas. When in solution, it releases hydrogen ions. When in solution, a base releases hydroxyl ions. On combining, positive hydrogen ions neutralise the negative charges of hydroxyl ions and form water and a salt. The following equation expresses this fact:

$$HCl + NaOH \rightarrow H_2O + NaCl.$$

VIII Give the word with the opposite meaning to:

(a) an acid
(b) positively
(c) lengthens
(d) loosens
(e) hind
(f) internal
(g) upper
(h) soluble
(i) visible
(j) attract

Unit 38 Diffusion

Vocabulary
according to concentration even perfume

A

The molecules of a substance are *moving about all the time without stopping*. Imagine a crowd of people leaving a football match. They are all trying to get away as quickly as *they can*, and this is *just/what the molecules of a gas do*. They try to spread out, or in other words, to diffuse.

5 *The spreading out of molecules* can be *shown* by opening a bottle of perfume in a room. The molecules of the perfume are *pushed very close together* inside the bottle. As soon as you open the bottle, they start to *spread out* into the air, and you can smell the perfume at the other end of the room. *The smell becomes stronger as long as the bottle is left open.*

10 *In the end*, when *the molecules are spread out evenly* inside and outside the bottle, the spreading out of molecules *stops happening*. *When we say this we mean* that both the room and the bottle *hold an even* mixture of perfume and air.

According to the *laws* of spreading out of molecules, the process *goes* 15 *on without stopping* until the concentration of perfume molecules is *the same all over* the room; in other words, until *the molecules have come to a condition where they are in balance*, and *do not change any more after this*, if conditions *stay without changing*.

B

The molecules of a substance are (1)in continual motion. Imagine a crowd of people leaving a football match. They are all trying to get away as quickly as (2)possible, and this is (3)exactly (4)how the molecules of a gas behave. They try to spread out, or in other words, to diffuse.

5 (5)Diffusion can be (6)demonstrated by opening a bottle of perfume in a room. The molecules of the perfume are (7)concentrated inside the bottle. Immediately the bottle is opened, they start to (8)diffuse into the air, and the perfume can be smelt at the other end of the room. (9)The longer the bottle is left open, the stronger the smell becomes.

(10)Finally, when (11)there is an equal concentration of molecules inside and outside the bottle, (5)diffusion (12)ceases. (13)By this is meant that both the room and the bottle (14)contain a uniform mixture of perfume and air.

According to the (15)principles of (5)diffusion, the process (16)con-
15 tinues unceasingly until the concentration of perfume molecules is (17)equal in all parts of the room; in other words, until (18)a state of equilibrium has been reached, and (19)there is no further change, if conditions (20)remain constant.

Exercise 1 Find the way in which the words and phrases italicised in Text A are expressed in Text B:

1 moving about all the time without stopping
2 they can
3 just
4 what the molecules of a gas do
5 The spreading out of molecules
6 shown
7 pushed very close together
8 spread out
9 The smell becomes stronger as long as the bottle is left open
10 In the end
11 the molecules are spread out evenly
12 stops happening
13 When we say this we mean
14 hold an even
15 laws
16 goes on without stopping
17 the same all over
18 the molecules have come to a condition where they are in balance
19 do not change any more after this
20 stay without changing

Exercise 2 Rewrite this passage using passive forms, without expressing the agent. (The subjects of the passive sentences are italicised):

We can compare *the movement of the molecules of a gas* to a crowd of people leaving a football match. We can compared *their rapid spreading out* to the behaviour of gas molecules. We call *this process* diffusion. We can demonstrate *diffusion* by opening a bottle of perfume in a room. The molecules start to diffuse as soon as we open *the bottle*. We can smell *the perfume* almost immediately at the other end of the room. Diffusion continues until it has reached *a state of equilibrium*, or until conditions change.

Exercise 3 Rewrite this passage, using more 'scientific' words or phrases to express the ideas in the parts italicised:

Gas molecules *move about all the time without stopping*. They *spread out/ easily and quickly*, as can be *shown* by opening a bottle of perfume. *As soon as it is opened*, the perfume can be smelt at a distance, and *the smell becomes stronger as long as the bottle is left open*. *In the end*, when both the room and the bottle *have in them an even* mixture of perfume and air, *the spreading out/stops happening*. If conditions *stay the same*, there is no *more change after this*.

Exercise 4 When two actions depend on each other, or change according to each other, the idea is often expressed as in (b):

(a) *The speed* of the settling depends on, or changes according to, *the size* of the particles.

(b) *The smaller* the particles, *the slower* they settle.

Note that the verb 'to be' can be left out in this type of sentence. Rewrite these sentences in the same way as in example (b):

1 (a) The speed of settling depends on the size of the particles.
  (b) The bigger—— the quicker——.

2 (a) If the particles are extremely small, they require a more powerful microscope to see them.
  (b) The smaller____ the more powerful the____.

3 (a) The speed of dissolving sugar depends on the heat of the water.
  (b) The hotter____ the faster____.

4 (a) The speed of the movement of the particles depends on the speed of the stirring.
  (b) The faster____ the faster____.

5 (a) If an atom is very complex, its nucleus is larger.
  (b) The more complex____ the larger____.

6 (a) The speed of sound changes according to the density of the medium through which it travels.
  (b) The denser____ the faster____.

7 (a) The amount of resistance offered by a copper wire depends on its thinness.
  (b) The thinner____ the greater____.

8 (a) The amount of resistance offered by a copper wire depends on its thickness.
  (b) The thicker____ the less____.

9 (a) The focal length of a convex lens decreases as its thickness increases.
  (b) The thicker____ the shorter____.

10 (a) A substance is a better insulator according to how much air it encloses.
  (b) The more____ the better____.

11 (a) The smell becomes stronger according to how long the bottle is left open.
  (b) The longer____ the stronger____.

12 (a) The speed of diffusion increases as the heat of the gas increases.
  (b) The hotter____ the faster____.

13 (a) The frequency of a wave becomes lower as the wave gets longer.
  (b) The longer____ the lower____.

14 (a) The frequency of a wave becomes higher as the wave gets shorter.
  (b) The shorter____ the higher____.

Exercise 5 Here are some sentences to be matched. Find the sentence on the right which expresses the same idea as the one on the left and combine them using BY THIS IS MEANT THAT:

1 Substances which enclose air are good insulators.
2 In a vacuum heat can be transferred by radiation only.
3 Free nitrogen cannot be used by green plants to build their tissues.
4 A plane mirror reflects an identical image of which the sides are reversed.
5 All waves possess frequency, amplitude and length.
6 No sound can be heard in a vacuum.
7 The frog is an amphibian.
8 The earth rotates round the sun in an elliptical orbit.
9 The body excretes waste material.
10 Eventually there will be an equal concentration of molecules.
11 The particles in a colloid are submicroscopic.

(a) They cannot be seen under an ordinary microscope.
(b) A medium must exist through which the waves can travel to the ear.
(c) A uniform mixture will be formed.
(d) It is not an equal distance away at all times.
(e) They do not conduct heat readily.
(f) A vacuum will prevent heat loss by conduction and convection.
(g) It eliminates substances which it cannot use.
(h) It begins life in water and completes it on land.
(i) It is laterally inverted but otherwise the same.
(j) They can assimilate it only when it is combined with other elements.
(k) They all have three characteristics in common.

Exercise 6 Answer these questions without referring to the Texts. (Do not number your answers, but write them in complete sentences as a paragraph, which will then be a summary of the Texts):

1 What are in continual motion?
2 How do gas molecules behave?
3 What is this process known as?
4 How can it be demonstrated?
5 When do the perfume molecules start to diffuse?
6 Where can the perfume be smelt?
7 When does diffusion cease?
8 What is another way of saying this?

Exercise 7 Questions for further discussion:

1 The oceans have derived their salt from the land. Why, then, isn't sea-water always much saltier near the land than in the middle of the ocean?

2 You might expect that there would be much more carbon dioxide in the atmosphere over the middle of the Sahara desert, where there are neither plants nor animals within hundreds of miles. However, this is not the case. Why not?

3 When something has burnt in the kitchen, you can smell it in a very short time all over the house, even if the doors are closed. Why is this?

Vocabulary

a desert

Exercise 8 Suggestions for further activities:

1 Put two lumps of sugar in a glass of hot water but do not stir. Observe what happens in time. Explain.

2 Give to a friend a number of closed bottles, which he will open in any order. Ask him to stand in a far corner of the room, while you stand in the middle of the room with your back to him. The closed bottles contain substances with a fairly strong smell, e.g. one of petrol, one of perfume, one of moth-balls, one of ammonia. Ask him to open any one of the bottles, without telling you which. Measure the time it takes for you to (a) notice a smell, and (b) recognise the smell. Which of the substances did you recognise in the shortest time?

Unit 39 Permeability

Vocabulary

cell special

A

The *word* 'permeability' means *being able* to *let something pass through*. If a material *lets water pass through*, it is said to be permeable to water. If a material will not let water pass through, it is said to be impermeable to water.

5 *Why do we wear* a special coat when it rains? Of what is the special coat made? It is *usually* made of rubber or plastic, *because* a thin layer of rubber or plastic will not let water pass through. In other words, rubber and plastic are impermeable to water. We should get wet if we wore a cloth coat in the rain, because cloth will let water pass through. In other 10 words, cloth is permeable to water.

We find thin layers of other impermeable materials in nature, too: e.g. the skin, which is impermeable to water. By this is meant that the skin will not let water pass through.

We know that all *plant and animal bodies/are made of* cells. Cells *are* 15 *like* little boxes *with liquid inside them*. The very thin *walls of the cells* are known as membranes, and they are permeable to water. But when the water *has something dissolved in it*, the membrane always lets the water pass through, *but* may not let the *dissolved substance* pass through. *When this is true*, the membrane is said to be *half*-permeable.

B

The (1)term 'permeability' means (2)the ability to (3)allow something to penetrate. If a material (4)allows water to penetrate, it is said to be permeable to water. If a material will not (4)allow water to penetrate, it is said to be impermeable to water.

5 (5)What is the reason for wearing a special coat when it rains? Of what is the special coat made? (6)As a general rule, it is made of rubber or plastic, (7)due to the fact that a thin layer of rubber or plastic will not (4)allow water to penetrate. In other words, rubber and plastic are impermeable to water. We should get wet if we wore a cloth coat in the
10 rain, (7)due to the fact that cloth will (4)allow water to penetrate. In other words, cloth is permeable to water.

Thin layers of other impermeable materials are found in nature, too: e.g. the skin, which is impermeable to water. By this is meant that the skin will not (4)allow water to penetrate.

15 It is known that all (8)living organisms (9)consist of cells. Cells (10)can be compared to little boxes (11)containing fluid. The very thin (12)cell walls are known as membranes, and they are permeable to water. But when the water (13)contains a solute, (14)although the membrane is always permeable to water, it may be impermeable to the (15)solute.
20 (16)In this case, the membrane is said to be (17)semi-permeable.

Exercise 1 Find the way in which the words and phrases italicised in Text A are expressed in Text B:

1 word
2 being able
3 let something pass through
4 lets water pass through
5 Why do we wear
6 usually
7 because
8 plant and animal bodies
9 are made of
10 are like
11 with liquid inside them
12 walls of the cells
13 has something dissolved in it
14 but
15 dissolved substance
16 When this is true
17 half-

Exercise 2 Rewrite this paragraph, using passive forms. You will then have summarised the Texts. (The subjects of the passive sentences are italicised):

If a material allows *water* to penetrate, we say that *the material* is permeable to water. If the material does not allow *water* to penetrate, we say that *the material* is impermeable to water. We find *thin layers of impermeable materials* in nature. We know that *all living organisms* consist of cells. We call *the cell walls* membranes, and they are always permeable to water. But the water may contain *a solute*. If the membrane is impermeable to this solute, we say that *it* is semi-permeable.

Exercise 3 ALTHOUGH is frequently used instead of 'but' in combining two ideas in one sentence. Notice that ALTHOUGH occupies a different place in the sentence:

(a) The sun appears to travel round the earth, *but* this is not really the case.
(b) *Although* the sun appears to travel round the earth, this is not really the case.

Rewrite these sentences, replacing 'but' with ALTHOUGH, as in example (b):

1 Things can be kept hot for long periods in a vacuum flask, but a little heat is always lost through the cork.
2 Rays from the sun warm everything they touch, but the rays themselves are quite cold.
3 A single plant produces thousands of seeds during one season, but very few of them survive.
4 The image in a plane mirror is identical, but it is laterally inverted.
5 The waves are moving outwards in ever-increasing circles, but the water itself does not travel.
6 The young tadpole breathes by means of gills, but the frog breathes by means of lungs.

7 The frog begins its life in water, but it completes it on land.
8 The large intestine is part of the alimentary canal, but digestion of food does not take place there.
9 The particles of a suspension can be seen under a microscope, but those of a colloid cannot.
10 Diffusion continues unceasingly, but it may take a long time for the molecules to reach a state of equilibrium.
11 A layer of plastic may be very thin, but it is impermeable to water.
12 A bean seed would begin to develop normally in the dark, but it would not survive for long.
13 A membrane may be permeable to some solutes, but it may be impermeable to others.
14 Animal bodies cannot assimilate free nitrogen, but they are unable to combine it with other elements.

Exercise 4 WHAT IS THE REASON FOR . . . ? is often used instead of 'Why . . . ?':

(a) *Why do we wear* a special coat when it rains?
(b) *What is the reason for wearing* a special coat when it rains?

Answer these questions, beginning each answer with:
The reason for —— is that ——.

1 What is the reason for the importance of oxygen?
2 What is the reason for the importance of nitrogen?
3 What is the reason for the importance of the nitrogen-fixing bacteria in the soil?
4 What is the reason for making a vacuum between the glass walls of a vacuum flask?
5 What is the reason for silvering these walls?
6 What is the reason for being able to see lightning before hearing thunder?
7 What is the reason for the absence of sound in a vacuum?
8 What is the reason for the chemical formula for atmospheric oxygen being O_2 and not simply O?
9 What is the reason for the atom being electrically neutral?
10 What is the reason for being able to smell perfume immediately on opening the bottle?

Exercise 5 Answer these questions without referring to the Texts:

1 What does the term 'permeability' mean?
2 Give an example of a material which is permeable to water, and of another which is impermeable to water.
3 What do all living organisms consist of?

4 To what may cells be compared?
5 What is a membrane?
6 Is a membrane permeable or impermeable to water?
7 What is meant by the term 'solute'?
8 What do you understand by the term 'a semi-permeable membrane'?

Exercise 6 Questions for further discussion:

1 If a substance is impermeable to water, it is called 'water-proof'. What do we mean when we say that something is:
(a) light-proof? (c) dust-proof? (e) thief-proof?
(b) rain-proof? (d) sound-proof? (f) shock-proof?
2 If a joint does not allow water to penetrate or escape, we say that it is 'water-tight'. What do we mean by the terms (a) 'air-tight'? (b)'light-tight'?
3 Give examples of things which are air-tight or light-tight.
4 What are the solutes in a glass of drinking water?

Exercise 7 Suggestions for further activities:

1 Take two pieces of the same kind of paper, and rub a little oil onto one of them. Drop some water onto each piece of paper. Into which piece does the water penetrate? What difference has the oil made to the permeability of the paper?
2 Take a piece of very thin skin from an onion and examine it under the microscope. The cells should be clearly visible. Draw what you see.

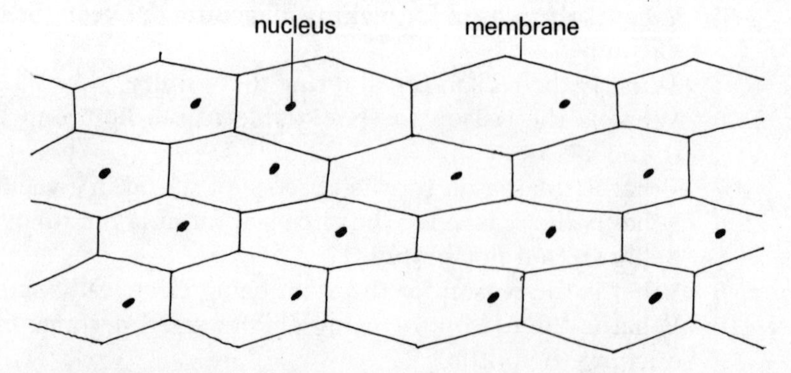

Diagram showing cells of thin onion skin under the microscope

3 Cut a raw potato in half and make a fairly large hollow in the middle as shown in the diagram. Stand the potato in a shallow dish of water and put some dry salt in the hollow. Leave it for two to three hours. Then notice how water has accumulated in the hollow where you put the salt. Where has this water come from? Is there more or less water in the dish now? Is the water in the hollow salty? Is the water in the dish also salty? (You will learn more about this in Unit 40).

Diagram of experiment

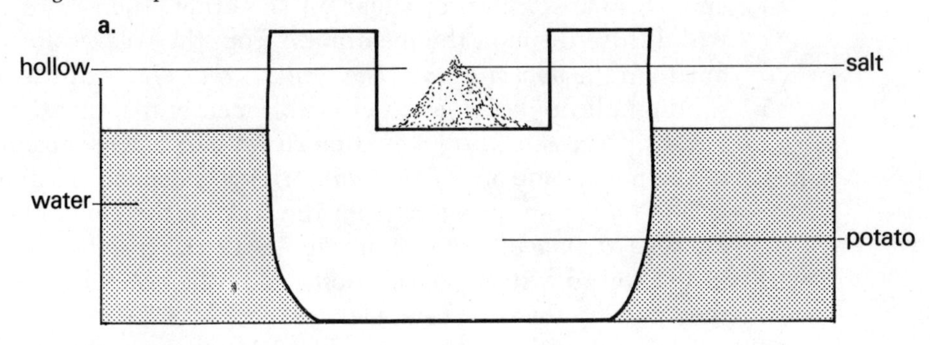

b. After 2-3 hours

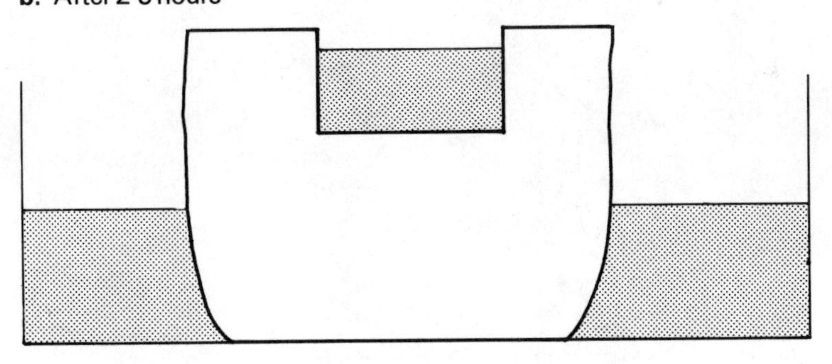

Unit 40 Osmosis

A

If a semi-permeable membrane has a *weak* solution of sugar on one side and a *strong* solution of sugar on the other, then *more* water molecules will diffuse through the membrane from the weak solution to the strong one than the *opposite way*. *The result is that* water diffuses from the weak
5 solution, through the semi-permeable membrane, into the strong solution.

This process is known as osmosis. When both solutions on either side of the membrane are of *the same strength*, osmosis *stops happening*.

The *taking up* of water from the soil by the roots of plants is a good *example to show* osmosis. Water molecules diffuse from the weak solution
10 in the soil to a stronger solution inside the *cells of the roots*. From here, they pass up the stem to the leaves.

All *plant and animal bodies* depend on osmosis to *carry/food materials* and waste materials through the cell walls. Osmosis is one of the most important *things that happen in nature*, and many life processes depend
15 upon it.

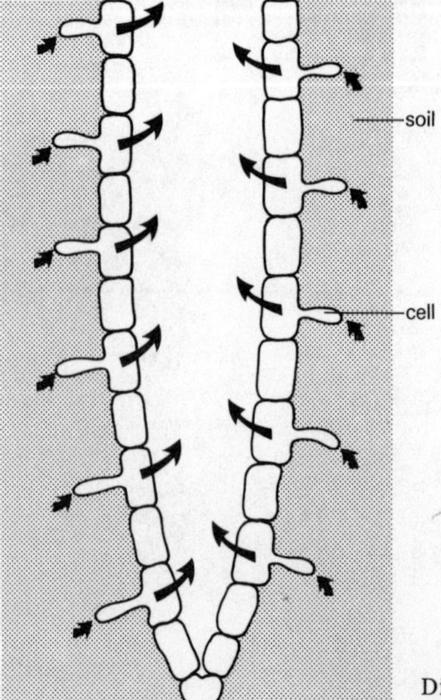

Diagram of root-hair (enlarged) absorbing water from the soil

B

If a semi-permeable membrane has a (1)dilute solution of sugar on one side and a (2)concentrated solution of sugar on the other, then (3)a greater number of water molecules will diffuse through the membrane from the (1)dilute solution to the (2)concentrated one than the (4)reverse. (5)In
5 effect, water diffuses from the (1)dilute solution, through the semi-permeable membrane, into the (2)concentrated solution.

This process is known as osmosis. When both solutions on either side of the membrane are of (6)equal concentration, osmosis (7)ceases.

The (8)absorption of water from the soil by the roots of plants is a good
10 (9)illustration of osmosis. Water molecules diffuse from the (1)dilute solution in the soil to a more (2)concentrated solution inside the (10) root cells. From here, they pass up the stem to the leaves.

All (11)living organisms depend on osmosis to (12)transport (13)nutrients and waste materials through the cell walls. Osmosis is one of the most
15 important (14)phenomena, and many life processes depend upon it.

Exercise 1 Find the way in which the words and phrases italicised in Text A are expressed in Text B:

1 weak
2 strong
3 more
4 opposite way
5 The result is that
6 the same strength
7 stops happening
8 taking up
9 example to show
10 cells of the roots
11 plant and animal bodies
12 carry
13 food materials
14 things that happen in nature

Exercise 2 Rewrite these phrases as compound nouns, as in example (b):

(a) *walls of cells* (b) *cell walls*

1 cells of roots
2 a solution of sugar
3 roots of plants
4 a process of life
5 organisms of plants
6 absorption of heat
7 impulses of nerves
8 transmission of energy
9 cells of animals
10 a layer of plastic
11 a coat of cloth
12 bodies of animals
13 cells of leaves
14 transport of nutrients
15 molecules of water
16 rays of light
17 nutrition of plants
18 an atom of hydrogen

Exercise 3 Compare these ways of expressing the same ideas:

In everyday speech or writing:	In Science:
(1a) to be different	(1b) *to differ*
(2a) to change	(2b) *to vary*
(3a) depending on	(3b) *according to*

(4a) The velocity of sound *changes depending on* the medium.
(4b) The velocity of sound *varies according to* the medium.

Rewrite these sentences as in examples (b):

1 The boiling point of different liquids is different.
2 The freezing point of different liquids is different.
3 The boiling point of a liquid changes depending on the pressure of the surrounding atmosphere.
4 The rate of expansion of gases, liquids and solids is different.
5 Water is different from most other liquids because it expands on solidifying.
6 The state of a substance changes depending on its temperature.
7 The temperature of the air will be different at different times of the day.
8 The climate in most places is different depending on the season.
9 The amount of rainfall on an area is different depending on the season.
10 The rate of heat transfer by conduction is different in different solids.
11 The method of heat transfer in solids and fluids is different.
12 The rate of development of plants will be different depending on their environment.
13 Methods of seed dispersal are different for different varieties of plant.
14 The focal length of a convex lens changes depending on its thickness.
15 The length of light waves changes depending on their frequency.

16 The frequencies of waves of light and sound are different.
17 The velocity of sound changes depending on the medium through which it travels.
18 The velocity of light and that of sound are different.
19 A colloid and a suspension are different depending on the size of the suspended particles.
20 The atom of one element is different from that of any other.

Exercise 4 Rewrite this passage, using passive forms. You will then have summarised the Texts. (The subjects of the passive sentences are italicised):

The semi-permeable membrane allows *water molecules* to pass through but will not allow *the sugar molecules* to do so. In osmosis, the semi-permeable membrane allows *the water molecules* to diffuse from the dilute to the concentrated solution. Roots of plants absorb *water and its solutes* from the soil in this way. They absorb *water from the dilute solution in the soil* into the concentrated solution inside the root cells. Living organisms depend upon *osmosis* for many of their life processes.

Exercise 5 Read and rewrite this passage, using a single word in place of each phrase italicised. (One small *change in word order only is necessary):

If a material allows water to *pass through,* it is said to be permeable to water. A material which does not, is said to be impermeable to water. All *plant and animal bodies *that live/are made* of cells, *which are like* tiny boxes *and their* walls are known as membranes. These are permeable to water but not always to every *substance dissolved in it. This kind of* membrane is said to be semi-permeable.

If a semi-permeable membrane lies between a *very weak* and a *very strong liquid with a substance dissolved in it,* more water molecules diffuse from the *very weak* to the *very strong* than the *opposite way.* This process is known as osmosis, and it does not *stop happening* until the concentrations of molecules on both sides of the membrane are *the same.*

A good *way of showing (with an example of)* osmosis is the *taking up* of water from the soil by the *roots of plants,* which *have in them* a more concentrated solution than that in the soil. It is also by osmosis that *food materials* are transported in plant and animal cells, and this is one of the most important *things that happen in nature* on which life processes depend.

-ABILITY/-IBILITY

Adjectives ending in -ABLE/-IBLE may form a noun ending with -ABILITY/-IBILITY:

permeable — permeability
visible — visibility

They also form their negatives by putting IN-/IM-/UN- in front:

impermeable — impermeability
invisible — invisibility

Exercise 6 Complete the table below as shown in the examples above:

ADJECTIVE	NOUN	NEGATIVE ADJ.	NEGATIVE NOUN
permeable	permeability	impermeable	impermeability
visible	visibility	invisible	invisibility
variable			
penetrable			
soluble			
combustible			
diffusible			
able			
digestible			
assimilable			
movable			
available			
absorbable			

Adjectives ending with -IC

Some adjectives are formed by adding -IC to the noun:

volcano — volcanic

Notice the difference in pronunciation, thus: The stress moves forwards to the syllable just before the -IC ending, e.g. MAG-net – mag-NET-ic.

Exercise 7 Read and write the adjectives formed from these nouns, as in the above examples:

1 ion	8 atmosphere	15 metal
2 atom	9 dynamo	16 microscope
3 magnet	10 organ	17 carbon
4 state	11 sulphur	18 protoplasm
5 volcano	12 acid	19 alcohol
6 electron	13 base	
7 mercury	14 metre	

Nouns ending with -SIS, -SE, change to -TIC in the adjective, e.g. ellipse – elliptic.

Read and write the adjectives formed from these nouns:

osmosis – osmotic
synthesis – synthetic

Exercise 8 Answer these questions without referring to the Texts:

1 What is the difference between the two sugar solutions on either side of the membrane?
2 Which way will the greater number of water molecules diffuse?
3 What is this process of diffusion called?
4 When does it cease?
5 Give a good illustration of osmosis.
6 Which solution is more concentrated, that in the root cells or that in the water in the soil?
7 Where does the absorbed water travel from the root cells?
8 What do living organisms depend on osmosis to do?

Exercise 9 Questions for further discussion:

1 Why can't salt-water fish live in fresh water, or fresh-water fish live in salt water?
2 Why is it unwise for a ship-wrecked sailor to drink sea-water when he can't find fresh water?
3 Why should fertilisers be used only in very dilute solutions?

Vocabulary

fresh water   ship-wrecked

Exercise 10 Suggestions for further activities:

1 Water a healthy (but unwanted) plant with salt solution for a few days. The plant will die. Why?
2 Water a healthy (but unwanted) plant with a concentrated sugar solution for a few days. The plant will die. Why?

Revision Exercises IX (Units 38–40)

I Give the meaning in your own language of these words and phrases:

1 concentration
2 result
3 cell
4 according to
5 special
6 perfume

II Explain the meaning of:

1 in continual motion
2 in constant conditions
3 in effect
4 a dilute solution
5 as a general rule
6 equal concentration of molecules
7 permeable to water
8 transport of nutrients
9 a state of equilibrium
10 the principles of diffusion
11 immediately on opening the bottle

III Give ONE word meaning:

1 wall of a cell
2 in the end
3 food materials
4 without stopping
5 the same all over
6 stops happening
7 which allows something to penetrate

IV Answer these questions without referring to the Texts:

1 When is a material said to be (a) permeable to water? (b) impermeable to water? (c) semi-permeable?
2 What is the difference between a dilute and a concentrated sugar solution?
3 When would the diffusion of molecules from an open bottle of perfume cease, in constant conditions?
4 Give an illustration of the process of osmosis in nature.
5 Give an example of a material which is impermeable to water.
6 What do all living organisms consist of?
7 Is the membrane of a cell permeable or impermeable to water? How do you know?
8 When does osmosis cease, in constant conditions?

V Find the correct word with which to complete each of these sentences:

1 When a state of equilibrium has been reached, diffusion —— .
(a) continues (b) changes (c) ceases (d) begins

2 If a material allows water to penetrate, it is said to be —— to water.
(a) permeable (b) impermeable (c) semi-permeable (d) perfumed

3 Nutrients are transported to the cells of a living organism by a process known as —— .
(a) digestion (b) assimilation (c) secretion (d) osmosis

4 Thin cell walls are known as —— .
(a) bacteria (b) membranes (c) phenomena (d) nutrients

5 The absorption of water from the soil by plant roots is a good example of —— .
(a) metamorphosis (b) osmosis (c) permeability (d) equilibrium

6 The molecules of a substance are in continual —— .
(a) reaction (b) diffusion (c) solution (d) motion

7 Cells can be —— to little boxes containing fluid.
(a) allowed (b) composed (c) equalled (d) compared

8 Water diffuses from the —— solution through the semi-permeable membrane to the concentrated solution.
(a) dilute (b) uniform (c) constant (d) even

Vocabulary

Brackets indicate that the word or phrase occurs under Questions for further discussion.

according to
(amber)
ammonia
angle
anus
axes (pl.)
axis

(baking powder)
balance
a bar
to behave
the blow
brain

carbon
cell
characteristic
a charge
chemistry
clear
(clover)
to collect
compact
compactness
(compass)
concentration

(to deaden)
(a desert)
(detergent)
(diet)

to digest
digestion
direction
to disturb
dye
dynamo

ear-drum
the east
(echo)
element
equation
(equator)
even
exactly

(fertiliser)
(fibre)
flour
the flow
force
(fresh water)
frog
(fun-fair)

gill
(glue)

hammering
(helmet)

insect

intestine
iron filings

jelly
juices

lightning
liver
lizard

to mention
microscope
to multiply
muscle

nail varnish
negative
nerve
neutral
nitrogen
nourishment

opposite

pancreas
parallel
path
perfume
planet
positive
protein
purple

relative
(reptile)
resistance

to set (sun)
to shave
(ship-wrecked)
(siren)
smooth
sound
sour
special
sphere
spot
to sprinkle
steel
to stir
stomach
to swallow

tadpole
thunder
tilted
tissue
(toothpaste)

upside down

to vibrate
vibration
(vinegar)

the west

Glossary

ability	being able
able	can
to absorb	to take up, to take in
absorption	(the) taking up, taking in
to accelerate	to speed up
acceleration	(the) speeding up
to act as	to behave like
to achieve	to do
(in) addition to	as well as
to adhere	to stick
adult	fully developed
to affect	to change in some way
aid	help
air resistance	the push of the air against
alimentary canal	long tube from mouth to anus
(an) alkali	a base which is soluble in water
to allow	to let
although	but
ampere	unit of measurement of electric current
amphibian	beginning life in water and ending it on land
amplitude	distance from highest to lowest point of wave
and so on	and so it continues
appearance	look
to appear to	to seem to
approximately	about, more or less
as a general rule	usually
to assimilate	to take up and use
at a distance	far away
at three months	when it is three months old
at high velocities	when things are going very fast
atmospheric	of the atmosphere
at right angles (to)	at an angle of 90° (to)
to attract	to pull towards
attraction	(the) pull towards
available	present, there to be used

base	bottom
(a) base	substance possessing opposite properties to acid
to behave	to do, to act
being	is
by means of	by using
by this is meant	when we say this we mean
to calculate	to do a sum to find out
(the) case	true
(in that) case	if/when this is true
to cause	to be the reason why
to cease	to stop happening
cell wall	wall of a cell
centre	middle
certain	(1) some, (2) known
circular	round, like a circle
charged	loaded
chemical formula	written like this in chemistry
to chew	to crush with the teeth
colloid	dispersion of submicroscopic particles in a fluid
colloidal	in or of a colloid
combination	(the) joining together
to combine	to join together
common	often found
(in) common	each having the same, sharing the same
to compare	to say is like
complex	complicated
to be composed of	to be made up of
concave	thinner at the middle than at the edges
to concentrate	to gather together, to collect
concentrated	(1) pushed together, (2) collected (3) strong
concentration	(1) collection (2) having large number of molecules
consequently	so, therefore
to consider	to think about
considerably	much
to consist of	to be made of
constant	unchanging, un- changed, the same
constant	unchanging, unchanged, the same
to contain	to have in it, to have inside
continual	without stopping
to continue	to go on

to converge	to bend towards each other
converging (lens)	convex (lens)
convex	thicker in the middle than at the edges
crest	highest point
current	(1) movement of a fluid, (2) flow of electricity
to decompose	to break down into small parts
to demonstrate	to show (how it acts or works)
density	compactness
to derive	to get
to differ	to be different
to diffuse	to spread out
diffusion	(the) spreading out (of molecules)
dilute	weak
to dislodge	to knock out of its place
dispersion	(the) scattering
(at a) distance	far away
distant	far away
to diverge	to bend away from each other
diverging (lens)	concave (lens)
due to the fact that	because
(the) duration of	for how long
(in) effect	the result is
electric(al)	of electricity
electrically	in electricity
ellipse	an oval
elliptical	oval-shaped
engine	machine
to enlarge	to make larger, to get larger
equal	the same
equilibrium	balance
ever-	always
(is) evident	can be seen
exactly	just
to excrete	to throw out from the body
to exhaust	to use up, to finish
to exist	to be
explained previously	we have explained before
to express	to say
external	on the outside

(in) fact	really
to feed/fed on	to eat/ate
finally	in the end
fixed	unchanging
fluid	liquid or gas
focal length	distance from lens to focal point
focal point	spot where rays are concentrated
fore-legs	front legs
(the) former	the thing mentioned first
formula	short way of writing
for this purpose	to do this
frequency	number of waves per second
friction	rubbing
further	more
(no) further	no more after this
gastric	of or in the stomach
(as a) general rule	usually
to generate	to make (electric power)
given	certain, measured
globule	small ball
gravitational	of gravity
(a) greater number	more
(the) greater part	most
gullet	muscular tube leading to the stomach
having	with, which has
high density	very compact
(at) high velocities	when things are going very fast
hind-legs	back legs
identical	exactly the same
i.e.	that is, that means
if it is to	if it is going to
if it were not for	if there were no
illustration	example, picture
image	picture
immediately	at once, as soon as
impermeable	will not allow to penetrate
impulse	message
in addition to	as well as
to increase	to become more

to indicate	to show (by pointing)
to induce	to produce without touching
in fact	really
ion	electrically charged atom
ionic	of ions
in other words	or we can say
insoluble	will not dissolve in water
in solution	dissolved in a liquid
to interpret	to give a meaning
in that case	if/when that is true
in this way	like this
inverted	upside down
invisible	cannot be seen
jelly-like	like jelly
laterally	sideways
laterally inverted	the sides are reversed, turned the other way
(the) latter	the thing mentioned second
leading (to)	going (to)
to lengthen	to grow longer
to lie	to be found
like	the same
lining	thin inner layer
litmus paper	specially dyed paper
living organisms	plant and animal bodies
location (of)	the place (where)
long-sighted	cannot see near things clearly
low density	not very compact
to be magnetised	to be made into a magnet
to magnify	to make it look bigger
magnifying glass	a glass which makes things look bigger
to maintain	to keep
(has) many uses	is useful in many ways
maximum	the greatest
by means of	by using
(a) medium	material to use (as a carrier)
membrane	thin cell wall
merely	only
metamorphosis	a great change
microscopic	can be seen under the microscope

(has a) milky appearance	looks like milk
mirror-image	picture reflected in a mirror
to modify	to change slightly
to moisten	to wet, to make wet
motion	movement
much	a lot
muscular	having muscles
natural phenomena	things which happen in nature
to neutralise	to make neutral
nitrogenous	of nitrogen
non-	which is not, which does not
nutrients	food materials
nutrition	feeding
(an) object	a thing
to obtain	to get
to offer	to give
ohm	unit measuring electrical resistance
on . . . ing	when . . .
opening	hole
orbit	path around something
to orbit	to travel round in an orbit
organism	living body
(in) other words	or we can say
owing to the fact that	because of
in all parts	all over
particle	tiny piece (of a solid)
to penetrate	to pass through
per	in one, in every
permeability	(the) ability to allow to penetrate
permeable	will allow to penetrate
phenomenon/ phenomena (pl.)	thing/s that happen/s (in nature)
to place	to put
plane	flat
to possess	to have
(as) possible	as they can
potential	amount of electricity (measured in volts)

powerful	strong
previously	before this
principle	law
to produce	to make, to manufacture
projection	a thing that sticks up
property	characteristic
protoplasm	living tissue
(for this) purpose	to do this
quantity	amount
(a) radical	group of atoms which behaves as a single atom
to react	to do something, to have something happen
reaction	happening
(for this) reason	this is why
to receive	to get
to refer	to look back
to reflect	to throw back
regardless of	no matter
to release	to set free
to remain	to stay
to remove	to take away
to repel	to push away
to require	to need
to resemble	to look like
resistance	(the) pushing against
(the) reverse	the opposite way
(at) right-angles	at an angle of 90°
root cells	cells of roots
to rotate	to turn round, to go round
rotation	(the) turning round, going round
saliva	juices in the mouth
to secrete	to send out
section	part
semi-	half-
to settle	to sink to the bottom
to shorten	to make shorter, to become shorter
similar	like
similarly (to)	just as, in the same way
simultaneously	at exactly the same time
since	because

single	only one
slightly	a little
so it is with	it is also like this with
soluble	will dissolve
solute	substance which has been dissolved
(in) solution	dissolved in a liquid
(and) so on	and so it continues
source	a few
space	(1) emptiness (2) place beyond earth's atmosphere
spawn	frog's eggs
spherical mirror	mirror which is part of a sphere
stage	time
state	condition, form
to strike/struck	to hit/hit
submicroscopic	cannot be seen with ordinary microscope
such	this kind of
such as	like
sufficient	enough
surrounding	all round, round about
to suspend	to hang about
suspension	dispersion of microscopic particles in a fluid
to take place	to happen, to occur
term	word, phrase
terminal	end
that is to say	this means
thus	so, in this way
transmission	(the) sending to another point
to transmit	to send to another point
to transport	to carry
type	kind, sort
unable	cannot
unceasingly	without stopping
to undergo	to have happen to it
undigested	not digested
uniform	the same all over, even all through
unlike	not the same
unsupported	not held up
(has many) uses	is useful in many ways
varieties	kinds, sorts

to vary	to differ, to be different
velocity	speed
(and) vice versa	also true if said the other way round
visible	can be seen
volt	unit measuring electric potential
waste	which cannot be used
water-insect	insect which lives in water
water-plant	plant which lives in water
wave-length	distance from one crest of a wave to the next
(in this) way	like this
where	when we write
whereas	but unlike this
with	which has, having
with the aid of	with the help of, by using

Notes for Teachers' Guidance on Questions for Further Discussion

Unit 23 Falling Bodies (Exercise 8)

1 (a) A trick question. They both weigh the same.
(b) A pound of lead has a higher density.
(c) A pound of lead would.
2 Its shape reduces air resistance.
3 When travelling at higher velocities, i.e. on a motor cycle, because greater air resistance is met with at higher velocities.
4 In order to reduce the effects of air resistance.

Unit 24 Nitrogen Fixation (Exercise 5)

1 Nitrogen. Essential for plant growth, not usually in sufficient amounts in soils repeatedly used for crops such as cereals, fruit and vegetables.
2 Clover is one of family whose roots contain bacteria which fix atmospheric nitrogen. Ploughing it back into soil after it has combined nitrogen in this way is a simple, cheap and effective way of enriching soil for further crops.
3 Decaying protein of animals supplies soil with nitrogen, which is used by plant to manufacture plant proteins.

Unit 25 Light: (1) Lenses (Exercise 5)

1 No. Any transparent substance will do, e.g. plastic, crystal, diamond, water. Only the shape is important.
2 Yes. Because water in plastic bag assumes shape of a converging lens (i.e. thicker in the middle).
3 They act as converging lenses, concentrating sun's rays onto dry grass, leaves, etc., thus setting them alight on a sunny day.
4 If sun's rays are concentrated on curtains, they may catch fire since water-filled glass acts as a converging lens.

Unit 26 Light: (2) Reflection in mirrors (Exercise 5)

1 Slightly concave.
2 Some are slightly convex to reflect an image covering wider field than a plane mirror does. Make object appear a little smaller.

3 Many different shapes. They bend light waves so as to produce comic distortions in image. Part of image may be magnified, part diminished, etc. They make people appear very fat, very thin, very short-legged, etc.

4 Because cheap mirrors are not absolutely plane, they distort image by bending the light rays. Distant objects appear more distorted because light rays travelling further suffer a greater divergence.

Unit 27 Wave Motion (Exercise 5)

1 No. Hat will merely bob up and down, unless driven by wind or current.

2 Sound waves: echoes, sonar, megaphones.
Light waves: reflections in mirrors, water, polished surfaces; reflectors used on cars, bicycles, advertisements or road signs (cats' eyes, etc.); radar, telescopes, microscopes, periscopes, etc.
Radio waves: television waves; reflected back to earth from layers in the ionosphere, radar, telstar.
Water waves: bounce back when they strike rocks or shore.

Unit 28 Sound (Exercise 6)

1 A trick question, really. Strickly speaking, no. Sound waves, i.e. disturbance of medium, would exist; but since no living thing is present, nerve impulses to brain, which interprets the waves as sound, are not present.

2 Reflection of sound waves from a surface. Any large, fairly plane surface will act as a reflector – cliff-face, wall, dome, high ceiling, mountainside, etc.

3 Yes. The shorter, the higher; the longer, the lower.

4 Short pipe gives higher note.

5 Because moon has no atmosphere, there is no medium in which sound waves can travel. Radio waves can travel in a vacuum.

6 To protect ear-drum from excessive vibration caused by jet engine. This can be painful and possibly damaging, and deafness may result.

7 No. They have a special organ which detects vibrations of sound waves in water.

8 With an echo sounder (sonar). Instrument sends out sound waves downwards from ship. Waves are reflected from sea-floor. Knowing the velocity of sound in water (approximately 4,800 feet per second), depth can be calculated by timing echo.

9 (Specially in the pre-radar days.) To find, by echo-timing, the distance to any existing iceberg.

10 Because of the absence of sound-absorbing bodies, like wood or furniture, cloth in furnishings, human bodies, clothes, carpets, etc. Bare surfaces act as sound reflectors.

11 Count the number of seconds between your seeing the lightning and hearing the thunder. Multiply this by 330m. E.g., if you count 5 seconds between the flash and the crash, the storm is $5 \times 330\text{m} = 1650\text{m}$. (or approximately 1 mile) away.

12 Any substance which absorbs sound waves, e.g. wood, fibreglass, cork, plastic, etc.

Unit 29 Magnetism (Exercise 7)

1 Various. E.g. two small plastic dolls, each containing a magnet. As they are pushed together, they are either attracted or repelled and can be made to 'dance'. E.g. small plastic car containing a piece of iron can be made to travel about on a piece of cardboard by moving a magnet under it.

2 It consists of a magnetised needle, balanced on a point so that it can swing. The needle always comes to rest in a north–south direction, because of the attraction of earth's poles.

3 Hold compass horizontally, and wait for needle to come to rest. This shows you in which direction north is, and you can then decide in which direction you should go.

4 Place the compass near either pole of magnet. If needle is attracted to pole, it indicates that this is the south pole of magnet. If repelled, it must be the north pole of magnet.

Unit 30 Movement of the Earth (Exercise 6)

1 24 hours.

2 $365\frac{1}{4}$ days.

3 The $\frac{1}{4}$ day is ignored for three years and accumulates as one whole day (29th February) once every four years.

4 Sun's rays are more direct, also travel through slightly less atmosphere (which absorbs heat).

5 Approximately 93 million miles.

6 Warm and sunny, Australia being in the southern hemisphere. Tilt of earth's axis as it rotates round sun causes change of season at different times in northern and southern hemispheres.

Unit 31 Atoms and Molecules (Exercise 8)

1 Less than 100.

2 (a) Oxygen, by far. Approximately 50% of the biosphere is oxygen.
(b) Other common elements are silicon, aluminium, iron and carbon.

3 (a) Compound (NaCl)
(b) Compound (CO_2)
(c) Compound (N + various elements)
(d) Compound (H_2O)
(e) Element (O_2)
(f) Element (H_2)
(g) Compound (carbon + oxygen + nitrogen + hydrogen and sometimes + sulphur or other elements)
(h) Element (N_2)
(i) Compound (NH_3)
(j) Compound ($KMnO_4$)

4 The remainder would fit into a small, but extremely heavy, shoe-box.

5 No, man has made a few.

Unit 32 Static Electricity (Exercise 6)

1 Positive and negative charges build up through friction of water vapour in clouds and air molecules, and attract each other – the more negative are attracted to the more positive. The flash is the sudden and rapid movement of electrons.

2 Through friction caused by body movement, charges of static electricity build up, and the negatively charged atoms are attracted to the positively charged.

3 Amber is a fossilized resin. Its Greek name is *elektron*, which is where the word 'electricity' comes from. It is thought that in about 600B.C. Thales first discovered static electricity by rubbing a piece of amber on flannel, and finding that it then attracted small bits of wood and straw.

Unit 33 Electricity (Exercise 5)

1 Driving all kinds of engines, trains, motors, etc. Refrigeration, heating, lighting.

2 Mountainous countries with high rainfall or snow, e.g. Norway, Sweden, Switzerland, can develop hydro-electricity to make supplies cheap. Countries with little coal, flat or dry lands, must produce electricity by more expensive means.

3 From the wheel, which is turned by the pedal, which is operated by the foot.

4 Accumulators, as used in motor vehicles, submarines, etc. Dry-cell batteries commonly used in torches, radios, tape-recorders, etc.

5 To increase current, make wire thicker to decrease resistance, or increase the potential. To decrease current, make wire thinner to increase resistance, or decrease the potential.
6 Thunder and lightning storms, and other examples mentioned under 'Static Electricity'. St Elmo's fire, Aurora borealis, certain fish (e.g. electric eel, electric cat-fish, electric ray) which can emit electric charges.

Unit 34 Life History of a Frog (Exercise 6)

1 Cold-blooded. All reptiles (e.g. lizards, snakes, tortoises) and all fish.
2 Early life on earth began in water; gradually water creatures evolved into land creatures, while some remained as water creatures. The frog represents a half-way stage between these two, since it starts life as a water creature, and ends it as a land creature.
3 Large numbers of eggs do not survive; they may be destroyed by predators, drought, lack of food, etc. To ensure survival of a few, thousands of eggs are produced every season.
4 Salamander, newt, toad.
5 (a) Hibernate. (b) Aestivate.
6 *Differences*
Only frog undergoes metamorphosis.
Respiration: Fish – gills only.
Lizard – lungs only.
Frog – first gills, then lungs.
Skin: Lizard and fish have scales, frog does not.
Habitat: Fish – entirely aquatic, all temperatures.
Lizard – land animal, mostly warm places.
Frog – amphibious, not very cold places.
Fish have no feet.
Similarities
All are vertebrates, all are hairless, all are egg-laying through sexual reproduction, all have tongues and teeth (very few frogs are toothless). None have ears, feet are clawless. Most can change colour, most are carnivorous – largely insect-eating, larger fish mostly eat smaller fish.

Unit 35 Digestion in Humans (Exercise 8)

1 From it we obtain energy and heat. We grow and renew our tissues. Without it life would cease.
2 Animal bodies cannot absorb food in form in which it enters mouth. Digestion breaks food down by chemical means and makes it soluble.
3 Milk requires little digestion, yet contains all nourishment required

for growth. Babies' digestive system not developed enough to take other foods, but in a few months, their system becomes able to assimilate fruit and vegetables in a form which does not require chewing.

4 When liver functions are impaired through illness, juices necessary for digesting fats are insufficiently secreted. A low-fat diet reduces the work of the liver.

5 (a) Carnivorous – i.e. meat-eating.
(b) Herbivorous – i.e. grass-eating.
(c) Omnivorous – i.e. will eat almost anything.

6 Lion's front teeth are long and sharp for killing and for tearing raw meat, thick and flat at the back for crushing bones. Sheep's teeth are prominent and chisel-shaped at front for chopping grass, flat at the back for cud-chewing. Man has both sharp cutting teeth at front and flat crushing teeth at the back, to cope with his varied diet of meat, vegetables, fruit, nuts, cereals, etc., but his jaws are relatively weak.

7 Man eats both high- and low-energy content foods – protein fat and carbohydrate. Stomach stores food only temporarily. Sheep eat low-nutrient content food – grass – so large amounts must be consumed. Lion eats only high-energy concentration food – meat – and eats a large amount at one sitting. This provides enough energy for a longer period.

8 Humans cannot digest/assimilate cellulose, the main content of wood. His teeth and jaws are unsuitable. He doesn't like the taste.

9 Man cooks most of his food. Man often eats for pleasure. Man often has elaborate social rituals connected with eating and drinking.

Unit 36 Solutions, Suspensions and Colloids (Exercise 8)

1 (a) Suspension – soil particles in water.
(b) Solution of dyes, or suspension of chemicals or pigments in water.
(c) Suspension – dust particles in air.
(d) Suspension/solution – red and white blood cells suspended in lymph, which is a solution.
(e) Solution of sugar in water, suspension of particles of lemon.
(f) Solution – sugar, chemicals, etc., carbon dioxide for the fizz.
(g) Colloid – chemical substance in water.
(h) Solution – salts, gases, etc.
(i) Suspension – water droplets containing soluble chemicals in air, e.g. sulphur dioxide.
(j) Suspension – pigments in water.
(k) Suspension – water droplets in air.

2 The electron- or ultra-microscope.

Unit 37 **Acids, Bases and Salts (Exercise 7)**

1 Various answers.
2 Potassium hydroxide (KOH) in soft soaps, and sodium hydroxide (NaOH) in hard soaps.

Unit 38 **Diffusion (Exercise 7)**

1 It diffuses evenly throughout. (May be slightly saltier near coasts of hot regions owing to faster evaporation of water.)
2 By diffusion is spread evenly throughout earth's atmosphere.
3 By diffusion, smell spreads quickly through air.

Unit 39 **Permeability (Exercise 6)**

1 (a) Does not allow light to penetrate.
(b) Does not allow rain to penetrate.
(c) Does not allow dust to penetrate.
(d) Does not allow sound to penetrate.
(e) Does not allow a thief to penetrate.
(f) Does not allow shock to affect it.
2 (a) Joined so tightly that air cannot pass.
(b) Joined so tightly that light cannot pass.
3 Space helmet, diving helmet; camera, wrapping on photographic films.
4 Air, carbon dioxide and minerals.

Unit 40 **Osmosis (Exercise 9)**

1 Rate of osmosis is dependent on the strength of the two solutions involved – one inside, one outside fish. Fish metabolism is adjusted to water in which it lives. If this changes too much, the metabolism cannot cope with different rate of osmosis required.
2 By the principle of osmosis, sea water cannot be absorbed by the body. Sailor would die, through loss of body water, however much sea water he drank. Body water would pass into digestive system and be excreted instead of assimilated. Body would them become dehydrated.
3 By the principle of osmosis, water passes from weak to strong solution, consequently, if fertiliser solution were stronger than solution inside root, water would pass from plant to soil, instead of from soil to plant. Plant would probably die of dehydration.